Prototyping Using Other Transactions

Case Studies for the Acquisition Community

LAUREN A. MAYER, MARK V. ARENA, FRANK CAMM, JONATHAN P. WONG,
GABRIEL LESNICK, SARAH SOLIMAN, EDWARD FERNANDEZ,
PHILLIP CARTER, GORDON T. LEE

Prepared for the United States Air Force
Approved for public release; distribution unlimited

RAND PROJECT AIR FORCE

For more information on this publication, visit www.rand.org/t/RR4417

Library of Congress Cataloging-in-Publication Data is available for this publication.
ISBN: 978-1-9774-0537-1

Published by the RAND Corporation, Santa Monica, Calif.
© Copyright 2020 RAND Corporation
RAND® is a registered trademark.

Support RAND
Make a tax-deductible charitable contribution at
www.rand.org/giving/contribute

www.rand.org

Preface

In the National Defense Authorization Act for 2016, the U.S. Congress made permanent the legal authority for the U.S. Department of Defense (DoD) to pursue other transactions for prototype projects (OTs) to engage in prototype development efforts that are not subject to many of the federal laws and regulations governing contracts, grants, or cooperative agreements as described in the Federal Acquisition Regulation (FAR) and the Defense FAR Supplement (10 U.S.C. 2371b). Congress's aim was to provide DoD with the means to gain access to the most innovative, creative technology available from nontraditional defense contractors to, ultimately, deliver lethality to the warfighter.

OTs have been touted for their potential to speed up the prototype development process, engage nontraditional and small business contractors, and take fuller advantage of innovative business models compared with FAR-based transactions. However, the use of OTs may also result in the U.S. Air Force being more susceptible to undisciplined use when less oversight and formalism exists.

To help Air Force leaders understand how the department is using the authority, whether its purported benefits are actually being achieved, and how the Air Force can improve the effectiveness of OTs, RAND Project AIR FORCE (PAF) was asked to review the Air Force's recent experience with using OTs for prototype projects.

The research reported here was sponsored by the Deputy Assistant Secretary of the Air Force for Acquisition (Contracting) and conducted within the Resource Management Program of RAND PAF as part of a fiscal year 2019 project, "Accelerating Modernization and Innovation: Use of Other Transactions for Prototype Projects Authority."

This report should be useful both to members of the Air Force acquisition community and to acquisition leadership. Other services' DoD agencies may also find the report useful, as many perspectives and lessons offered throughout the report may be generalizable to the larger defense acquisition community. While we provide a short background about OTs, we assume that readers of this report have a basic understanding of OT authority and the defense acquisition process.

RAND Project AIR FORCE

PAF, a division of the RAND Corporation, is the U.S. Air Force's federally funded research and development center for studies and analyses. PAF provides the Air Force with independent analyses of policy alternatives affecting the development, employment, combat readiness, and support of current and future air, space, and cyber forces. Research is conducted in four programs: Strategy and Doctrine; Force Modernization and Employment; Manpower, Personnel, and Training; and Resource Management. The research reported here was prepared under contract FA7014-16-D-1000.

Additional information about PAF is available on our website: http://www.rand.org/paf

This report documents work originally shared with the U.S. Air Force on September 11, 2019. The draft report, issued on September 27, 2019, was reviewed by formal peer reviewers and U.S. Air Force subject-matter experts.

Contents

Figures

Tables

Summary

Issue

RAND Project AIR FORCE was asked to review the Air Force's recent experience using the authority for other transactions for prototype projects (OTs)—which allows the U.S. Department of Defense (DoD) to develop prototypes outside of the traditional federal laws and regulations governing contracts, grants, or cooperative agreements. The aim is to help U.S. Air Force leaders understand how the service is using the authority, gauge whether they are achieving their purported benefits, and envision ways the service can improve their effectiveness.

Approach

Using literature reviews, interviews, and case studies of seven OT projects funded by the Air Force, the research team identified lessons and observations to understand each OT's policy context, purpose, and decisionmaking. The team sought to answer the following questions: How has the Air Force used OTs? What outcomes are associated with the use of OTs? What lessons from the use of OTs might be helpful to acquisition professionals? What challenges exist with using OTs?

Observations

How Has the Air Force Used OTs?

- Air Force use of OTs increased between January 2016 and December 2018 by nearly fourfold (see Figure S.1). While one program (National Security Space Launch) dominates this dollar total, the number of OT projects awarded also follows a similar trend.
- The Air Force has made diverse use of OTs. For example, our case studies included efforts for experimental tests, weapon and IT systems, and physical and business processes.

What Outcomes Are Associated with the Use of OTs?

- Our case study research suggested that personnel using OTs believe the authority provides a number of flexibilities not inherent to Federal Acquisition Regulation (FAR) procurements.
  - They allow for more freedom to communicate with industry, tailor solicitations and agreements, and work under conditions acceptable to nontraditional firms.
  - The government can engage in a more commercial-like manner and conduct untraditional forms of market intelligence, thus attracting nontraditional sources.
- Cost-sharing, an important incentive tool, is less difficult through an OT in many cases.

Figure S.1. Total Air Force Funded OT Obligations by Year

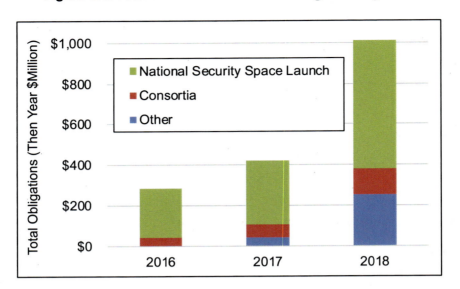

NOTE: RAND analysis of Federal Procurement Data System—Next Generation (FPDS-NG) data.

What Lessons from the Use of OTs Might Be Helpful to Acquisition Professionals?

- OTs are but one of many tools available to the Air Force for developing prototypes. Other mechanisms may also provide the necessary flexibilities in some cases. Government personnel need to determine the most effective mechanism for the problem.
- Effective OT teams balance OT flexibility with an appropriate level of discipline.
  - Even though not required for OTs, most case studies competed the OT to broaden the potential solution set and documented key decisions to provide internal traceability.
  - Early and frequent communication with all OT stakeholders can help to align expectations.
  - Tying milestone payments to key deliverables is an important incentive mechanism.
  - The principles behind the FAR are still good guidance for formulating OT agreements.

What Challenges Exist with Using OTs?

- While the Air Force has been able to leverage OT flexibility, a number of challenges remain.
  - Since few generalizable, prescriptive rules apply to OTs, compliance-based training methods are insufficient.
  - Establishing institutional OT knowledge is difficult, as Air Force OT experience is still developing, documentation is not required, and staff turnover is somewhat common.
  - Limited institutional knowledge coupled with OT statute ambiguity (e.g., what constitutes a prototype?) leads to some disagreements and rework.
  - Air Force contracting culture is inherently compliance-based and risk averse. What constitutes "success" is elusive. This results in discomfort with necessary risk-taking.
  - Agreements officers are chosen from a pool of the most experienced and creative contracting officers, potentially causing a less balanced overall acquisition workforce.

Recommendations

- *Adjust Air Force acquisition environment to provide relevant training, facilitate information sharing, and manage the OT workforce.* Training can be case-based, focusing on problem-solving. Sharing of OT lessons can be facilitated through more formal coordination, allowing for continuous learning. The OT workforce can be strategically managed to include formal mentoring programs and provide for broader experience (i.e., acquisition, technical, legal).

- *Adopt an Air Force culture that values taking calculated risks.* OT personnel need to be rewarded, not punished, for their willingness to take risks to accomplish the mission and for using sound judgment. To be successful, the culture shift must be adopted not only by senior leaders but also by all levels of management.

Acknowledgments

We thank Major General Cameron G. Holt for sponsoring this work and for supporting its execution. We also thank Jeanette Snyder for her assistance in carrying out this research. Many others in the U.S. Air Force, too numerous to mention by name, shared their insights with us and participated in our case study research. Additional discussions held with individuals from the Office of the Under Secretary of Defense for Acquisition and Sustainment, the Defense Acquisition University, the Defense Advanced Research Projects Agency, the Congressional Research Service, the Government Accountability Office, George Mason University Center for Government Contracting, the U.S. Army Contracting Command, an OT consortium management firm, and the Defense Innovation Unit provided valuable insights.

At RAND, we thank Phil Anton and Jeff Drezner for their analytic reviews, Jake McKeon and Maynard Holliday for helpful discussions, and Michelle Horner and David Richardson for their administrative assistance. We also thank Moshe Schwartz for his analytic review.

That we received help and insights from those acknowledged above should not be taken to imply that they concur with the views expressed in this report.

Abbreviations

ACC	Army Contracting Command
AFLCMC	Air Force Life-Cycle Management Center
AFRL	Air Force Research Laboratory
AMI	Applied Minds Inc.
AO	agreements officer
AOC	Air and Space Operations Center
AOR	agreements officer representative
ATI	Advanced Technology International
BAA	Broad Agency Announcement
C5	Consortium for Command, Control, and Communications in Cyberspace
CICA	Competition in Contracting Act
CO	contracting officer
COFC	Court of Federal Claims
CRS	Congressional Research Service
CSO	commercial solutions opening
C-UAS	counter–unmanned aerial systems
DARPA	Defense Advanced Research Project Agency
DCMA	Defense Contract Management Agency
DFARS	Defense Federal Acquisition Regulation Supplement
DIU	Defense Innovation Unit
DoD	Department of Defense
EELV	Evolved Expendable Launch Vehicle
FAR	Federal Acquisition Regulation
FBO	Federal Business Opportunities (a.k.a. FedBizOpps)
FPDS-NG	Federal Procurement Data System—Next Generation
GAO	Government Accountability Office
ID/IQ	Indefinite-Delivery, Indefinite-Quantity

IP	intellectual property
IT	information technology
LSA	Launch Service Agreement
NASA	National Aeronautics and Space Administration
NDAA	National Defense Authorization Act
NSSL	National Security Space Launch
OBAC	OPIR Battlespace Awareness Center
OPIR	Overhead Persistent Infrared
OSAI	Open System Acquisition Initiative
OSD	Office of the Secretary of Defense
OT	other transactions for prototype projects
OTA	Other Transactions Authority
PM	program manager
R&D	research and development
RFI	Request for Information
RFP	Request for Proposal
RPP	Request for Prototype Project
RPS	Rocket Propulsion System
SAF/AQC	Deputy Assistant Secretary of the Air Force for Acquisition (Contracting)
SBIR	Small Business Innovation Research
SDPE	Strategic Development Planning & Experimentation Office
SMC	Space and Missiles System Center
SOW	Statement of Work
SpaceX	Space Exploration Technologies Corp.
SpEC	Space Enterprise Consortium
TAP	Tools, Applications, and Processing

1. Introduction

Section 815 of the National Defense Authorization Act (NDAA) for 2016 permanently codified authority for the Department of Defense (DoD) to carry out other transactions for prototype projects (OTs).[1] OTs are a government agreement mechanism for developing prototypes that are not subject to many of the traditional federal laws and regulations governing contracts, grants, or cooperative agreements, as described in the Federal Acquisition Regulation (FAR) and Defense FAR Supplement (DFARS). The intent of this authority is to provide the DoD with the means to gain access to the most innovative, creative technology available from nontraditional defense contractors to, ultimately, deliver lethality to the warfighter when needed.

This and other recent legislation was intended to reinvigorate the department's use of this authority and has prompted numerous discussions surrounding its use. The advantages of OTs have been promoted by the Air Force (U.S. Air Force, n.d.) as well as the press (e.g., Maucione, 2017). These claims include those of (1) greater speed, flexibility, and accessibility of prototype projects; (2) promoting engagement of nontraditional and small business contractors; and (3) usefulness in designing and implementing innovative business models that may not otherwise be feasible or practical under FAR-based transactions.

However, the use of this authority may also result in the Air Force being more susceptible to various execution problems when less oversight and formalism exists. In fact, Congress has recently passed draft legislation for the 2020 NDAA requiring formal reporting on the use of OT agreements (Williams, 2019). Further, some DoD decisionmakers are concerned that the ability to engage in OTs is a transitory authority (Shelbourne, 2019).

Given the lack of a well-established track record on its recent use, we were asked by the Deputy Assistant Secretary of the Air Force for Acquisition (Contracting) (SAF/AQC) to gain a better understanding of the Air Force's recent experience with using OTs by addressing the following questions.

- How has the Air Force used OTs?
- What are the outcomes associated with the use of OTs?
- What are the enduring lessons from the use of OTs that might be helpful to acquisition professionals?

---

[1] The abbreviation "OT" in this document is used specifically for "other transactions for prototype projects." Most publications use this abbreviation to represent all types of other transactions (i.e., research and development; prototype projects). Given the narrow focus of the report on other transactions for prototype projects, use of the abbreviation specifically for prototype projects is meant to simplify its language.

- Is there potential to improve the effective use of OTs? If so, what changes in law, policy, or surveillance might be required?

The purpose of our research was to help the Air Force better understand how it is using OTs and whether the purported benefits are actually being achieved. It also should illuminate the challenges in implementing this agreement mechanism.

The scope of the research focused almost exclusively on other transactions for prototype projects and their follow-on production (10 U.S.C. 2371b). When we refer to OTs in this report, we refer solely to those for prototypes and their associated follow-on production. This report does not address other transactions for research (10 U.S.C. 2371).[2] Finally, while there are many other innovative procurement mechanisms available to the Air Force, such as 10 U.S.C 2373—*Procurement for Experimental Purposes*—and Section 804 of the NDAA for FY 2016—*Middle Tier Acquisition*—these are beyond the scope of this report. All references to OTs for the remainder of the report are to those authorized as part of 10 U.S.C. 2371b.

Chapter 2 begins by presenting a more complete definition of the OT authority, data on OT use by the Air Force since 2016, and a literature review of recent OT research and policy. Chapters 3 through 6 present observations from a series of cases studies of seven recent Air Force OT projects. Chapters 3 through 5 focus on observations and lessons that are useful to the government practitioner (e.g., agreements officers [AOs], program managers [PMs], requirements owners). Chapter 6 is more applicable to senior decisionmakers in the Air Force and DoD, first providing a brief summary of the insights gleaned from the acquisition community presented in the previous three chapters and then discussing some higher-level policy considerations. Chapter 7 summarizes our main conclusions.

Our report addresses the study questions bulleted on the previous page in discussions that cut through several chapters. Table 1.1 on the following page lists the study questions and the chapters where readers can find our explorations of them. We also return to the questions in reviewing our study conclusions in Chapter 7.

A series of appendices provides further detail for the interested reader. Appendix A describes our case study methodology in detail. Appendix B provides some background and history for each of our case studies. Appendix C explores some of the legal issues related to bid protests of OT projects.

---

[2] For a detailed description of the distinction between the different types of other transaction authority, see the section on "Types of OTs" in DoD's *OT Guide* (OUSD[A&S], 2018).

Table 1.1. Study Questions and Chapter(s) Where They Are Discussed in Report

Study Question	Chapter(s) in Which Question Is Discussed						
	1	2	3	4	5	6	7
How has the Air Force used OTs?		✓				✓	✓
What are the outcomes associated with the use of OTs?		✓				✓	✓
What are the enduring lessons from the use of OTs that might be helpful to acquisition professionals?			✓	✓	✓		✓
Is there potential to improve the effective use of the OTs? If so, what changes in law, policy, or surveillance might be required?						✓	✓

2. Legislative Background and Recent Air Force Use of OTs

Understanding how the Air Force is using OTs requires first understanding the underlying motivations for their employment. In this chapter, we provide a more detailed definition of an OT and the motivation for the recent codification of DoD's OT authority. We summarize recent research and data that address whether OT use is achieving Congress's objectives and highlight existing research gaps. Finally, we provide a brief overview of our research approach, aimed at filling some of these gaps.

OT Definition

In its simplest form, an OT is a government agreement mechanism for prototypes that waives many traditional government contract regulations, including the FAR, its supplements and DFARS. OTs are typically defined by what they are not: they are not standard contracts, grants, or cooperative agreements (10 U.S.C. 2371b; OUSD[A&S], 2018).

While not required to comply with the FAR, OT projects are subject to a few conditions. First, they must be "directly relevant to enhancing mission effectiveness"[1] and, for DoD, published OSD policy provides the following prototype definition:

> addresses a proof of concept, model, reverse engineering to address obsolescence, pilot, novel application of commercial technologies for defense purposes, agile development activity, creation, design, development, demonstration of technical or operational utility, or combinations of the foregoing. A process, including a business process, may be the subject of a prototype project. (OSD, 2018)

Second, they must meet at least one of the following conditions:

> (A) There is at least one nontraditional defense contractor or nonprofit research institution participating to a significant extent in the prototype project.

> (B) All significant participants in the transaction other than the Federal Government are small businesses . . . or nontraditional defense contractors.

> (C) At least one third of the total cost of the prototype project is to be paid out of funds provided by sources other than other than the Federal Government.

> (D) The senior procurement executive for the agency determines in writing that exceptional circumstances justify the use of a transaction. (10 U.S.C. 2371b)

---

[1] The full language from the statute reads: "directly relevant to enhancing the mission effectiveness of military personnel and the supporting platforms, systems, components, or materials proposed to be acquired or developed by the Department of Defense, or to improvement of platforms, systems, components, or materials in use by the armed forces" (10 U.S.C. 2371b).

Further, OT projects with award values above $100 million and $500 million require higher-level approvals from the agency's senior procurement officer[2] and DoD, respectively. Finally, some laws still do apply to OTs, such as criminal laws, export controls, and the Civil Rights Act (Sidebottom, n.d.).

Other transaction authority for "advanced research projects" was first granted to the National Aeronautics and Space Administration (NASA) in 1958. It remained the only agency with the authority until 1989, when Congress enacted 10 U.S.C. 2371 granting the Defense Advanced Research Project Agency (DARPA) the same other transaction authority for research efforts. Authority to use other transactions for prototype projects did not begin until 1993 with the passage of Section 845 of the NDAA for FY 1994; this expanded the authority to include prototype projects and opened it up to all of DoD. Successfully completed OT projects could also be carried over to a follow-production OT award (10 U.S.C. 2371b). Although initially temporary and somewhat limited in scope, Congress permanently codified DoD's authority in 2016. Between 1993 and 2019, the statute has been repeatedly updated to clarify its intent, update its scope, and add oversight.[3]

DoD and the Air Force have issued very few policies to prescribe how the acquisition community must follow the OT statute because each agreement is intended to be structured to meet the needs of the program. However, given the ambiguity in the statute, DoD has issued a series of guidebooks to provide some clarification. A comprehensive guide was first issued in 2002, updated in 2017 after the statute was permanently codified, and finally reissued by the Defense Acquisition University in 2018 (OUSD[A&S], 2018).[4] This guide provides nonprescriptive guidance for the use of OTs. DoD is careful to clarify that it is "not a formal policy document" and encourages government personnel using OTs to be creative and innovative when carrying out an OT.[5]

Motivation for OT Authority

The purpose of the FAR is to ensure that government acquisition activities are executed to "deliver on a timely basis the best value product or service to the customer, while maintaining the public's trust and fulfilling public policy objectives" (FAR 1.102). Congress understands that these acquisition goals, at times, can result in the U.S. government's losing its technological

---

[2] For the Air Force, the senior procurement officer is the Assistant Secretary of the Air Force, Acquisition, Technology, & Logistics (SAF/AQ).

[3] The Congressional Research Service (CRS) recently published a full legislative history, and we direct the interested to Schwarz and Peters (2019).

[4] The newest *OT Guide* is meant to be a "living document" and iteratively refined as new lessons are learned.

[5] However, the Government Accountability Office (GAO) has used DoD's guidebook to support industry bid protest decisions.

and competitive edge in the global marketplace.[6] Other transaction authority was established to ensure that this does not occur. Congress first designed this authority to allow NASA enough freedom in its research and development (R&D) to compete in the international space race. The expansion and permanent codification of DoD's OT authority came at a time when Congress believed there was a need to gain a competitive edge against near-peer adversaries in technology advancements and cyberspace. Recent national trends show that R&D spending and talent has been shifting toward supporting the commercial sector and away from government (Campbell and Shirley, 2018). To compete with U.S. adversaries, DoD now realizes that it needs to attract business from a wider range of commercial firms to access emerging technologies important to military use. DoD is therefore strategically promoting competing for business in ways it has never done before (e.g., Hagel, 2014). OTs are one (of many) mechanisms oriented at helping DoD retain technological advantage through innovation and rapid experimentation as well as enabling small nontraditional firms to participate in more commercial-like transactions.

The Joint Explanatory Statement of the NDAA for FY 2016 provides the following rationale for expanding and permanently codifying DoD's OT authority:

> The conferees believe that the flexibility of the OTA [Other Transactions Authority] authorities . . . can make them attractive to firms and organizations that do not usually participate in government contracting due to the typical overhead burden and "one size fits all" rules. The conferees believe that expanded use of OTAs will support Department of Defense efforts to access new source[s] of technical innovation, such as Silicon Valley startup companies and small commercial firms. (U.S. House of Representatives, 2015)

With the passage of Section 815 of the FY 2016 NDAA, Congress provided DoD with flexibilities in the hopes to accomplish three goals: (1) to reinvigorate the use of OTs, (2) to attract and access nontraditional sources, and (3) to promote innovation.[7]

OTs Provide DoD with Flexibility

In addition to exempting OTs from many federal acquisition regulations, the OT statute is written to be intentionally vague. These two factors allow DoD to exercise a number of flexibilities and tailor the acquisition process to the specific circumstances of each OT (Halchin,

---

[6] As an example, consider views provided by the U.S. Senate Committee on Armed Services in the NDAA for FY 2018 report for Section 804 (Clarification of Purpose of Defense Acquisition), which amended the introduction to the DFARS:

> The committee notes that the Department of Defense is constantly forced to balance equities related to . . . defense and national security goals and broader national and public policy goals. The Department also struggles to align goals relative to improving the speed and response to threats with public transparency and fiscal stewardship. . . . The committee remains concerned that these balances and goals sometimes drive the Department into practices that drive up costs, slow down the acquisition process, and result in sub-optimal capabilities being developed and deployed to operational forces.

[7] We note that Congress's motivation for expanding and codifying OT authority may differ from the Air Force's motivation to use this authority. However, recent Air Force testimony (Roper, 2018) signals that the intent is similar.

2011; Schwarz and Peters, 2019).[8] For example, OTs are exempt from the Competition in Contracting Act (CICA) and only require competition to the "maximum extent practical" (10 U.S.C. 2371b). OTs can, therefore, be awarded as sole-source agreements[9] or use more innovative (and potentially simpler) source selection processes similar to, for example, Broad Agency Announcements (BAAs) (OUSD[A&S], 2018; Defense Innovation Unit [DIU], 2016). OT projects may also be awarded to consortia. OT consortia can take on many forms but generally are an organized group of members in which a lead entity coordinates and directs the consortia activities (Schwarz and Peters, 2019). Numerous prototype projects may then be solicited for and awarded to consortium members under the consortia OT (Boyd, 2018).

The flexibility of OTs extends to the development of award agreements. While contracts, grants, and cooperative agreements are strictly prescribed by the FAR and DFARS, OT agreements have virtually no requirements. Without the need to comply with the Bayh-Dole Act, AOs can negotiate intellectual property (IP) terms freely as they would in a commercial setting (Fike, 2009). The stringent cost accounting standards to comply with DoD's Cost Accounting System required of contractors by the FAR also do not apply to OTs. While some DoD regulations still apply, OTs essentially allow AOs to develop an agreement "from scratch" and to meet the needs of the specific problem and contractor.

Two further flexibilities afforded with OTs worth noting are the lack of restrictions on government communications with contractors and the near elimination of documentation requirements during the source selection and award process[10] as well as during execution of the OT project. These flexibilities allow for a more commercial-like setting in which to conduct business.

Finally, numerous other flexibilities make OTs unique to more traditional contracting mechanisms: advanced payments are allowed, recovery of government funds are permitted, materials submitted under a solicitation for an OT are exempt from the Freedom of Information Act for five years from the date on which the information is received by DoD, contract terms for Termination for Default or Termination for Convenience can be negotiated, and there is flexibility in payment methods, as well as milestones (Sidebottom, n.d.).

---

[8] Note that nonstatutory parts of the FAR may be waived by acquisition executives; therefore, the flexibility provided by an OT is (1) providing exemptions from the statutory parts of the FAR and (2) not requiring waivers for the nonstatutory parts. While the statutory sections of the FAR do also provide for a number of flexibilities, these more flexible options are often not exercised. See further discussion on this in Chapter 6.

[9] A sole-source follow-on production OT can only be awarded if the original prototype was both competed and successfully completed. See 10 U.S.C. 2371b(f)(2).

[10] While documentation is not required, an agency's decision to use its OT authority and award a sole-source follow-on OT project is susceptible to industry protest. Contemporaneous documentation of decision rationale for pursuing an OT project and designating it as "successfully completed" has proven to be a key component to agencies' justifying their position during industry protests to GAO or the Court of Federal Claims (COFC) (see Appendix C for further information on GAO and COFC jurisdiction and protest decisions).

The flexibility of OTs is meant to increase the attractiveness to nontraditional sources of doing business with the DoD. The substantial regulatory and compliance hurdles of the FAR create barriers for small businesses, such as technology start-up companies, to enter into government contracts. OTs lower such barriers to entry. Cost-accounting requirements, documentation, and other formal processes add effort and resources that small businesses may not have to spare. OTs may also appeal to larger firms because of the flexibilities with IP, which is often the most valuable asset for businesses with an innovative technology to offer (Kelly, 2018). Traditional government contractors can enjoy these flexibilities either by ensuring they use a significant percentage of nontraditional suppliers or by entering into cost-sharing agreements (i.e., paying for at least one-third of the total cost of the project) with the government.

However, DoD cannot attract nontraditional sources unless those sources are aware of OT project opportunities. Without many of the restrictions of the FAR, the government can move away from the traditional Request for Proposal (RFP) process to use new and innovative ways of publicizing these opportunities such as prize contests (OUSD[A&S], 2018). The Defense Innovation Unit is one specific way that DoD is currently publicizing opportunities. It offers brick-and-mortar spaces in Silicon Valley and other innovation hubs that focus specifically on accessing and attracting nontraditional sources from those areas (DIU, n.d.).

In all, OT flexibilities attract the commercial sector to do business with DoD by giving them "the ability to enter into innovative business arrangements or structures that would not be feasible or appropriate" under traditional contracts, grants, or cooperative agreements (OUSD[AT&L], 2017; Dunn, 2009). With the current locus of innovation being held by the private sector (Mitchell, 2017), OTs can shift some of that back toward the defense sector.

Recent Evidence of Meeting OT Goals

In the three years since Congress passed Section 815 of the FY 2016 NDAA, there has been very little research that reviews whether OTs are meeting their original objectives. We summarize this recent research below.

Reinvigorating Use

Congress expanded and permanently codified DoD's OT authority with the intention of reinvigorating its use. Recent data on OT use suggest that it has met this objective. While completely accurate data on OT use is not available,[11] overall trends show significant growth

---

[11] Reporting of OTs does not conform well to the FPDS-NG and therefore is not reported consistently (Schwarz and Peters, 2019).

since 2016. The contracting intelligence firm, Deltek, reports that OT spending across all agencies with authority (e.g., DoD, Department of Homeland Security, Department of Transportation) more than doubled from $2.1 billion in 2017 to $4.3 billion in 2018 (Mazmanian, 2019), with $3.4 billion of that 2018 total spent solely by DoD (Temin, 2019). Some observers project that DoD will double its 2018 spending to $7 billion by the end of 2019 (Cornillie, 2019). In 2018, about 8 percent of DoD research, development, testing, and evaluation funds were awarded through OTs (Temin, 2019).

Our analysis of Federal Procurement Data System—Next Generation (FPDS-NG) data between January 2016 and December 2018 shows that the Air Force has funded at least $1.7 billion in obligations.[12] Figure 2.1 illustrates Air Force–funded OT obligations by year. Spending increased 49 percent from 2016 to 2017 and nearly 140 percent from 2017 to 2018. Obligations are dominated by the two National Security Space Launch (NSSL) program OT projects, with nearly 70 percent of obligations over the three years. Obligations for OT projects awarded under consortia hold the second largest amount in 2016 and 2017, with

Figure 2.1. Total Air Force Funded OT Obligations by Year (Then-Year $Millions)

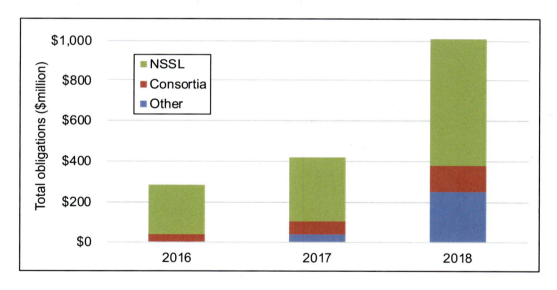

SOURCE: RAND analysis of FPDS-NG data.
NOTE: FPDS-NG data may underrepresent OT obligations (see Schwarz and Peters, 2019). Our review of FPDS-NG reporting of Air Force OT awards prior to 2016 suggests that these data are unreliable to use for comparison with 2016–2018 data.

non-NSSL standalone OT projects overtaking obligations through consortia in 2018. This trend may soon change, as the Space Enterprise Consortium (SpEC)—the Space and Missile Command's (SMC) OT consortium—is considering raising its $500 million ceiling to $12 billion

---

[12] Given the issues noted with FPDS-NG, we assume this is a lower-bound estimate. FPDS-NG data were pulled as of March 7, 2019, and includes Air Force–funded OTs awarded by other services (e.g., Army) and DIU.

over the next decade because of demand projections (Erwin, 2019). It is worth noting that while the Air Force funded all OT projects discussed here, a number were awarded through the Army, either through one of its OT consortia or in conjunction with the Defense Innovation Unit (DIU) (which until recently used the Army to contract all of its OTs). Air Force–funded, Army-awarded OTs between 2016 and 2018 amounted to approximately 13 percent of all Air Force–funded OT obligations.

FPDS-NG data supplemented with data obtained from SAF/AQC show that the number of OTs awarded also increased over time from 2016 through 2018, with a total of 27 OT agreements over the three-year period. Figure 2.2 presents the number of OT projects and agreements by year. In this case, multiple OT agreements may be awarded to different contractors under the same OT project (e.g., the NSSL Rocket Propulsion System [RPS] OT project was awarded to four contractors). The 27 OT agreements were awarded across 16 OT projects. In the figure, an OT consortium is counted as one OT project and one OT agreement. The Air Force awarded one OT consortium in 2016 and two in 2017. Each consortium has awarded multiple OT agreements (with those agreements being between a contractor and the consortium manager). Overall, the number of OTs awarded by the Air Force more than doubled between 2017 and 2018. For the 16 OT projects included in this data set, Figure 2.3 presents the number of OT projects by the total award value. Half of these projects included awards (sometimes across multiple agreements)

Figure 2.2. Number of Air Force OT Projects and Awards by Year

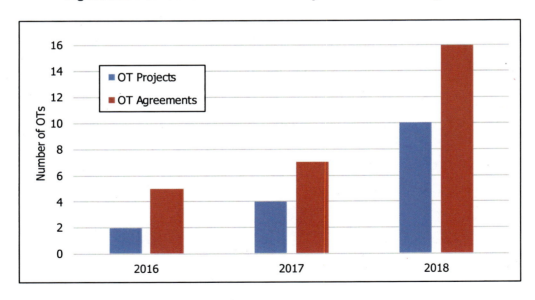

SOURCE: RAND analysis of FPDS-NG and SAF/AQC data.
NOTE: Multiple OT agreements may be awarded to different contractors under the same OT project. This chart counts an OT consortium as one OT project and one OT agreement. However, multiple agreements may be established between a contractor and a consortium manager under one OT consortium.

10

Figure 2.3. Number and Proportion of Air Force OT Projects by Award Amount

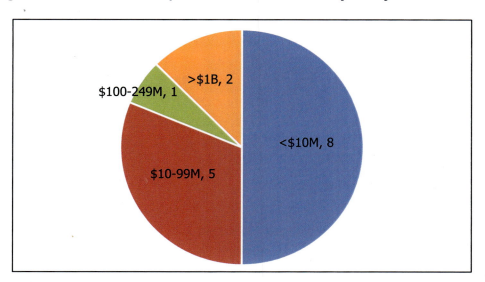

SOURCE: RAND analysis of FPDS-NG and SAF/AQC data.
NOTE: Multiple OT agreements may be awarded to different contractors under the same OT project. This chart counts an OT consortium as one OT project.

of less than $10 million, with another 31 percent (five projects) having a value under $100 million. Only a small number of projects had values greater than $100 million, in which two were the NSSL OTs.

The relative growth in OT use seen in these data is partially contrasted by a Government Accountability Office (GAO) report from 2016. It found that government agencies shied away from using OTs because they could not easily identify opportunities for use, and even when they did, their lack of experience meant that developing an OT agreement would require more time and effort than a traditional contract (GAO, 2016). Of course, as we note, the OT environment has significantly changed since the 2016 GAO study, potentially reducing the relevance of such findings. However, GAO (2016) did observe that government agency officials found creating and administering OT agreements especially challenging because of the lack of a standard structure based on policy guidelines. This is still the case today, though DoD's *OT Guide* (OUAS[A&S], 2018) attempts to provide some loose guidance on these aspects.[13]

Flexibility, Nontraditional Participation, and Innovation

GAO's (2016) interviews with government agency officials suggests that the acquisition community is finding that OTs do, in fact, provide a number of flexibilities. Officials believed that OTs were enabling the development of customized agreements that could better address contractors' concerns related to, for example, IP rights and cost-accounting compliance.

---

[13] Further guidance may be found in DIU's Commercial Solutions Opening guide (DIU, 2016).

One study cites that this flexibility is indeed attracting nontraditional sources to the DoD. Nontraditional contractors who spoke to the Congressional Research Service (CRS) stated that the existence of OTs would make them more likely to work with DoD (Schwarz and Peters, 2019).

The impact OTs have on DoD innovation is less clear. As Fike (2009) points out, "nontraditional defense participation by itself does not automatically lead to an advancement of technology for DoD." However, GAO (2016) reports that officials from most agencies it interviewed said that OTs "allowed them to carry out activities that they otherwise would not have been able to do."[14]

OT Timelines

While not the initial intent by Congress, OTs have also received the reputation for being faster than traditional FAR-based mechanisms (e.g., Maucione, 2017). Data that accurately compare the timeline between OTs and FAR contracts do not currently exist. However, we were able to use FPDS-NG and SAF/AQC data to analyze the OT timelines for a large subset (21 of 26) of the competed OT awards made between January 2016 and October 2018. Figure 2.4 illustrates the average OT timeline for these awards as well as the relationship between the individual OT timelines and their total award value. On average, OTs are awarded 188 days after the solicitation is released, with approximately three-quarters of that time attributed to postsolicitation response activities (e.g., evaluating proposals, selecting recipients, and negotiating the agreement). Overall, OT timelines appear to increase with larger total award amounts. Given that each OT is unique, we present three different OT "categories" in the figure: awards made for the NSSL, OT consortia, and other standalone OTs. Each category shows the general relationship of increasing timelines with increasing award values.

Once on contract, most (17 of the 27) Air Force OT agreements in our data set were executed with a period of performance of less than two years. The ten OTs with longer periods of performance were either OT consortia or related to the NSSL.

Research in Previous Periods of OT Usage

Although the DoD and Air Force usage has increased significantly since 2016, OT usage has been encouraged in the recent past, and research done then highlights many of the same issues considered here. Smith, Drezner, and Lachow (2002) concluded that OT usage in the late 1990s and early 2000s encouraged nontraditional participation, that OTs brought new and beneficial

---

[14] While these activities were not explicitly cited for DoD, the report does note that a DoD component was able to secure an agreement with a nontraditional contractor who sought special intellectual property provisions.

Figure 2.4. OT Timelines on Average and by Award Value (Then-Year $Million)

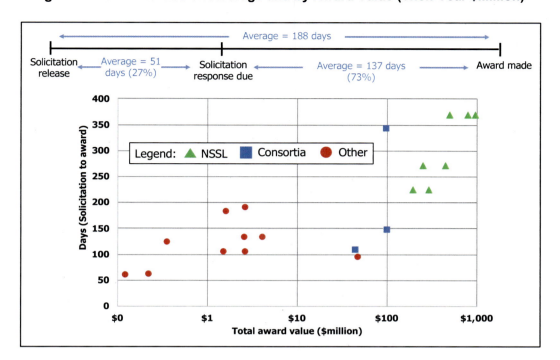

SOURCE: RAND analysis of FPDS-NG and SAF/AQC data.
NOTE: Total award value is presented using a log scale.

flexibility to contracting, and that risks resulting from the reduced oversight when using OTs were substantially outweighed by its benefits. Drezner and Leonard (2002) found that OTs facilitated the necessary working relationships to encourage collaborative prototyping efforts with industry but cautioned that the government needs to take particular care to exercise oversight. More recently, Pernin et al. (2012) found that the Army's use of OTs for its ambitious Future Combat Systems project was appropriate given the pathbreaking and unconventional nature of the program but that the agreements were not properly structured to balance flexibility with oversight. While the research presented in this report focuses on current Air Force usage of OTs, it is useful to know that previous research echoes issues, concerns, and observations about current OT usage.

Our Research Addresses Existing Gaps

Data provided in the previous section show that OT use has significantly increased since 2016. However, very little research has been undertaken to better understand how the authority is being used and how that use can be improved. Of the research that does exist, none explore the experience of those who actually implement OTs—the acquisition community.

Our research addresses these gaps by conducting semi-structured interviews with members of the acquisition community involved in OTs to understand how they are using them. We completed case studies of seven recent Air Force–funded OT projects initiated after 2016 (Table 2.1). These

Table 2.1. OT Case Studies

Name	Affiliation	Government Obligation	Period of Performance (Years)	Features
Counter-UAS	AFRL	$2.0M (in-kind)[a]	< 1	Cost share, experimental test
AOC Pathfinder/ Project Kessel Run	AFLCMC, Army, DIU	$21M	1	Prototype, information technology (IT)
	AFLCMC	$48M	1	Follow-on, IT
NSSL Rocket Propulsion Systems	SMC	$740M	3+	Cost share, system
Tetra project, Space Enterprise Consortium	SMC	$18M	1–2	Consortium project, system
TAP Lab and OBAC project, Consortium for Command, Control, and Communications in Cyberspace (C5)	SMC, Army	$2.1M	1–2	Consortium project, IT design
Engine manufacturing process	AFRL	$1.2M	1–2	Sole-source, physical process
Technology accelerator	AFRL, AFWERX	$2.5M	< 1	Business process

NOTE: AFLCMC: Air Force Life-Cycle Management Center; AFRL: Air Force Research Laboratory; AOC: Air and Space Operations Center; OBAC: Overhead Persistent Infrared (OPIR) Battlespace Awareness Center, TAP: Tools, Applications, and Processing.
[a] The C-UAS award was a zero-dollar transaction but developed an in-kind cost-sharing agreement.

cases range in affiliation, obligations, and period of performance as well as other features such as whether the OT used a cost-share, sole-source, or consortium project award. Our primary source of information for these cases came from interviews with the users associated with these OTs, including AOs, requirements holders, PMs, and consortium managers. We analyzed these interviews using a number of well-documented qualitative methods, which resulted in the lessons and observations documented throughout this report. A further discussion of our case study analysis methodology can be found in Appendix A. Each case study is explained in more detail in Appendix B.

We supported insights from our case study interviews with a detailed review of related OT documents (e.g., agreements, solicitation). Additional reviews of prior OT research, current OT guidance, policy and training, industry OT protests, and contract law provided a broader context in which to perform the case study analysis.

While our case studies represent a cross section of the OTs recently pursued by the Air Force, caution should be taken when generalizing our findings to all Air Force OT projects. During our research, we learned that each OT is quite unique, and practices associated with executing OTs are in an experimental phase. Therefore, it is important to frame these analyses as an early, systematic look at lessons being learned in a rapidly maturing field. For this reason, we refer to our findings as *perspectives*, with the intention that they create a benchmark for many of the lessons the Air Force has learned in its short history of using OTs.

Further, our goal in performing this research was principally to gather an organized summary of perspectives informed directly from our case study research. Therefore, it does not provide a comprehensive guide to using OTs. Rather, it focuses on the lessons that OT *users* believed to be important. For more comprehensive guidance on using OTs, we direct interested readers to DoD's *OT Guide* (OUSD[A&S], 2018). On a similar note, we present a number of perspectives that may either be helpful with FAR-based approaches or in other cases may differ from those in FAR-based approaches. If our case study participants explicitly noted that, we reproduce those perspectives here. Otherwise, a comparison with FAR-based contracting was considered out of scope for this work.

Additionally, our research does not offer conclusive insight into improper use of OTs or whether the use of OTs violates the principles of fairness and probity in traditional government contracting. Such concerns include the potential for poor management oversight and fiscal accountability, among others (Schwarz and Peters, 2019; Smith, Drezner, and Lachow, 2002; Drezner and Leonard, 2002). The focus of this research was on the experiences of OT users in the Air Force; our interviews and case studies did not focus on improper use or uncover any related insights, although we acknowledge that this is an important issue to consider.

Our research is framed around the premise that insights and lessons from the acquisition community's recent experience with OTs can (1) inform and improve the effective future use of OTs for the acquisition community and (2) inform actions taken by senior Air Force leadership to facilitate that effective use in a responsible way. The next three chapters provide those insights and lessons specifically focused for the acquisition community.

3. The OT Life Cycle: Phases, Overarching Characteristics, and Challenges

Our case studies illuminated a number of user perspectives that should be helpful to the government team (especially requirements holders[1] and AOs) when considering whether and how to use OTs. We provide these perspectives over the next three chapters.

Appendix B provides further details for how these perspectives relate to each of our case studies. In this chapter, we develop a framework for conceptualizing the OT environment. That is, we lay out the OT Life Cycle or the chronological process for implementing an OT. The phases of this life cycle were informed heavily by our case studies and supplemented with other OT-related information sources such as the DoD *OT Guide* (OUAS[A&S], 2018) and Air Force OT training materials. It represents a generalized process for implementing an OT. Given that every OT is unique, some of our case studies had slight deviations from the process explained below. We also discuss in this chapter some overarching considerations for OT users that apply across that life cycle. These were informed almost solely from analyses of our case studies. Chapters 4 and 5 provide considerations from the case studies organized around this OT Life Cycle. In Chapter 4, we discuss considerations for OT users during the initial phase of the life cycle (presource solicitation). Chapter 5 provides considerations for the remainder of the life cycle.

The OT Life Cycle

Across our case studies, we found that OT users follow a general process when implementing an OT. This process begins even before deciding to use an OT and continues until the OT has been executed and closed out. Many phases of the process are iterative and depend on previous decisions made in previous phases, and each one must be informed by the unique circumstances of the OT project.

The first step in the process is to define the problem. As discussed further below, this problem definition may be iteratively refined throughout the entire life cycle. With a problem defined, a conceptual strategy for soliciting sources and awarding the OT can be developed. This involves a number of interdependent processes. Depending on the situation, any one or combination of them can occur first. The requirements holder must identify funding for the project. They also must choose a contracting office to execute the project, and together this

---

[1] We have defined the term "requirements holder" for the purposes of this report. Given the flexibility of OTs, a number of organizations can perform a role similar to that of the user community or PM in an OT. These roles are further defined in a subsequent section of this chapter. To provide broadly applicable perspectives, we use this term throughout the report.

government team can decide on the appropriate award instrument (e.g., whether to use an OT) and perform market research. The interdependencies between these processes are many. This results from the inherent flexibility of OTs to rapidly adapt to changing circumstances. As an example, market research may indicate that only nontraditional sources are able to meet the need, in which case the use of an OT may be appropriate. However, as described later in the report, the funding stream may not align with its use due to color-of-money issues. The knowledge acquired during these activities may also result in iterations on the problem definition.

More market research may be necessary after the decision has been made to use an OT—for example, in determining whether and how to solicit sources. Finally, a solicitation can be released. Given the fewer restrictions on communications, feedback from and negotiations with offerers can occur throughout the solicitation process. This can lead to further iterations of the problem. Once final proposals have been received, the government team must evaluate offers and select those for potential OT awards. Prior to making the award, the agreement must be developed. Negotiations with the offerer may again result in iterations to the problem. Finally, the OT can be executed, modified over the course of the period of performance if necessary, and closed.

Figure 3.1 depicts this OT life cycle as four basic phases, beginning with defining the problem through award and execution. As shown in the figure, the phases don't always proceed in a strictly linear fashion. Defining the problem occurs throughout the life cycle and is therefore represented with feedback arrows at every other phase. It is also possible that iterations on the

Figure 3.1. The OT Life Cycle

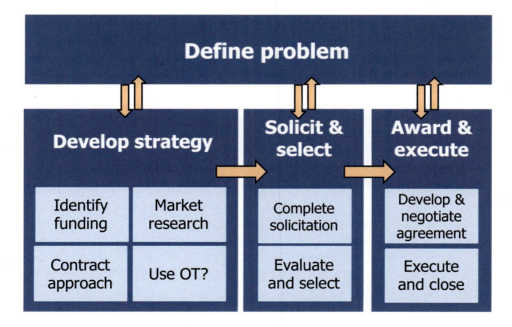

problem definition that occur in response to information gained during the solicit-and-select phase may require the government team to restart the process and refine the strategy.

The OT Government Team

Many government personnel are involved in the execution of an OT. Below we provide a brief generalized overview of the primary personnel and their roles and responsibilities with respect to OTs. Again, since every OT is unique, the personnel, roles, and responsibilities may differ slightly with each.

- Requirements holder: The requirements holder defines the problem. In the context of an OT, a requirements holder may be an operational user or, as in most of our case studies, be affiliated with a specific program office. Requirements holder personnel, such as the PM or the agreements officer representative (AOR),[2] are usually involved throughout the entire OT life cycle. For instance, these personnel help to seek out funding, perform market research, develop the technical sections of the solicitation and agreement, evaluate offers, and work with the contractor during OT execution.
- Agreements officer (AO): The AO is a contracting officer with a warrant to issue, administer, and terminate OT agreements.[3] AOs usually are not involved in the initial problem definition phase but may be involved in iterative refinements to the definition— for instance, to ensure the effort aligns with OT statute. Our case study research suggests that the AO and requirements holder need not come from the same office. Requirements holders can choose to use a contracting office outside of their organization (and even outside of their service). The AO works with the requirements holder to develop the strategy for and execute source solicitations and the OT award. They play a major role in the development of the terms and conditions for an OT agreement.
- Legal and policy personnel: Legal and policy personnel from the contracting office overseeing the OT's execution may also be involved at many stages during the OT life cycle. They provide guidance on complying with legal statutes and DoD policy relevant to OTs. Therefore, they review draft solicitations, agreements, and OT award decisions as well as help to determine agreement compliance once the OT is executed.

While not explicitly part of the government team, personnel from an OT project's funding source may be involved in a number of decisions during the OT life cycle. OT project funding is often

---

[2] AORs appear to be a role created specifically for OTs. While we are unaware of any official documentation that defines this role, the OT agreements we reviewed all had a designated AOR. In some instances, they are affiliated with a program office while others are affiliated with the operational user. One OT consortium included in our case studies defines the AOR role as "the individual designated by the Government on a per-Project basis to monitor all technical aspects and assist in administration of the specific Project" (Consortium Management Group, 2017). One case study participant stated that AOR responsibilities include monitoring performance, reviewing deliverables, and accepting deliverables, among others.

[3] DoD AOs need not be warranted contracting officers; that decision is delegated down to the individual services and agencies (OUSD[A&S], 2018). As of December 2018, the Air Force had not issued service-wide procedures but held the position that all Air Force AOs be warranted contracting officers (Snyder, 2018). At least one Air Force Command has issued official memorandum to this effect (Bullock, 2018).

received outside of the requirements holder's office, representing another stakeholder in the OT formation and execution process.

For each phase of the OT life cycle, we collected lessons, observations, and strategies from OT users. We identified a number of considerations that apply across the entire life cycle, which we discuss in the following sections.

The Rules of Good Contracting Still Apply

The OT process outlined previously is relatively similar to that used when implementing FAR-based transactions. Accordingly, it should not be surprising that many insights from our case studies can be summarized as standard best practice for any government contracting activity. That is, we found that effective use of OTs was facilitated by

- instituting early and frequent communication and coordination among the government team (e.g., AOs, requirements holders, legal, policy)
- defining the problem and developing solicitation requirements and the Statement of Work (SOW) by
  - gathering and consolidating viewpoints from relevant experts and mission partners
  - carefully balancing the language to be flexible enough to allow for innovation and learning, but specific enough to set clear objectives
  - considering the use of cost-sharing agreements that invest industry in positive project outcomes
- understanding that OTs are one mechanism in the contracting "toolbox," which should be used only when appropriate for the specific problem and circumstances
- engaging early with and developing a detailed understanding of the market
- developing OT agreements by
  - aligning payable milestones to key targets of the project to ensure contractor performance is associated with the government's project management expectations
  - considering any dependencies to proper project execution (e.g., project depends on another contract being successfully completed)
- maintaining situational awareness during OT project execution of business conditions outside the government's control that could result in problems, such as cost or schedule growth, and adjusting the OT agreement or plans proactively.

In addition to these more generic government contracting best practices, we also learned that many AOs still rely on FAR/DFARS constructs when they execute OTs. That is, the FAR embodies a number of commonsense principles that may be applied to the OT environment.

- *Promotion of competition:* While awarding a sole-source OT award has many fewer restrictions than in FAR-based acquisitions, soliciting multiple sources for an OT project may be especially important when trying to access nontraditional sources that can provide innovative solutions. The commercial market for OT project offerers is less well known to the DoD than that for traditional defense contractors. Without publicizing problems and holding a competition, the Air Force could be missing out on the best means of solving the problem. Receiving multiple offers for a solicitation may also allow the Air

Force additional leverage during award negotiations. Given the flexibility that OTs offer during agreement negotiations, the Air Force has the ability to conduct a competition that is more similar to that performed in the private sector.

- *Documentation of important decisions:* While almost no documentation is required during OT execution, we found that many government personnel still documented OT-related decisions, such as the technical evaluation of offers, negotiation memos, acquisition strategy, decisions to sole-source awards, and so on. Documentation was not lengthy, and many personnel commented that they used a "document as necessary" approach. Personnel provided us with numerous reasons as to why they document decisions.
    - Documentation creates a record. As turnover is likely in many government positions, maintaining a record of the rationale for decisions will allow for smoother transitions and retaining of institutional knowledge.
    - Documentation is still necessary for some Air Force–level reviews (e.g., legal review of awards). In this case, the documentation allows for all government personnel to ensure they are in concurrence and maintains situational awareness.
    - Documentation can help when industry bid protests have been lodged. While many decisions made during OT execution are not under GAO or Court of Federal Claims (COFC) jurisdiction, an agency's decision to use its OT authority and award a sole-source follow-on OT project is susceptible to industry protest. Contemporaneous documentation of decision rationale for pursuing an OT project and designating it as "successfully completed" may improve the Air Force's ability to justify its position during industry protests to GAO or COFC (see Appendix C for further information on GAO and COFC jurisdiction and protest decisions).

While many of these insights on the surface suggest that FAR-based contracting and OT execution are similar, the flexibility of OTs and intended goals of their use present unique challenges and require innovative strategies for their effective implementation.

Challenges in Implementing OTs Compared with More Traditional Approaches

Throughout our case study work, we identified a number of unique challenges that the government team faces when implementing an OT. These challenges, we heard, increase the likelihood of issues, such as disagreements, among OT stakeholders and rework. In this section, we provide a summary of those challenges.

- *OT knowledge is currently in a nascent stage.* While the Air Force has had the authority to use OTs for over two decades, very few individuals currently have OT experience. This includes not just the Air Force acquisition community but also the traditional and nontraditional contractors. Given this limited OT experience, the most effective and efficient ways to implement an OT are still being discovered.
- *There are few generalizable, prescriptive rules that can be applied to OTs.* Complicating the former challenge is the fact that the unique qualities of OTs result in very few rules that can be applied to the OT environment that are both generalizable and prescriptive in nature. This fact makes it difficult to train individuals using the more compliance-based methods often used for FAR-based contracting mechanisms. Our case study participants

20

stated that compared with the current methods of training in DoD, "on-the-job" training was the only effective way to learn how to implement an OT.

- *Compliance with the OT statute is ambiguous*. Part of the reason that there is a lack of a generalizable, prescriptive OT rule set is the intentionally sparse and ambiguous nature of the OT statute. The statute provides great freedoms to government teams working an OT, but with those freedoms, it also creates unanswered questions. The two most prominent questions we heard during our case studies were: (1) What constitutes a prototype OT, and (2) how should a successful OT completion in the OT agreement be defined?[4] The ambiguity in answering such questions can create disagreements among the different OT stakeholders (e.g., legal counsel, funders) and may result in rework and delays.

- *Institutional knowledge about OTs is difficult to maintain*. Partially as a result of the lack of a generalizable, prescriptive rule set, we learned that institutional knowledge about OTs is difficult to maintain. This includes both general knowledge about effective implementation as well as knowledge specific to an ongoing OT. The ability to maintain such knowledge is additionally complicated by the lack of requirements to document OT-related decisions. Without a "paper trail" for this information, institutional knowledge may be lost. Further, many of our case studies included instances of government personnel turnover, from AOs to legal counsel to AORs. This turnover led to some loss in institutional knowledge, sometimes resulting in disagreements, rework, and delays.

Communication

One last overarching theme that we heard across every case study was the importance of early and frequent communication across the entire OT life cycle. We identified four types of communication that are essential to effectively executing an OT.

1. Communication within the government team
2. Communication between the government team and contractors
3. Communication between the government team and other OT users
4. Communication between the government team and higher-level stakeholders.

In these roles, early and frequent communication can help to

- anticipate and mitigate potential disagreements or opposition, reducing iterations and rework,
- develop working relationships that provide for a smoother execution and increased collaboration, and
- facilitate the sharing of lessons and sustain institutional knowledge.

Communication increases in importance as the number of stakeholders involved in an OT grows (e.g., as a result of additional sources of OT project funding or the use of a consortium). It is also essential to managing many of the unique challenges of the OT environment, as described in the previous section.

---

[4] While OSD issued guidance to define a "prototype project" and "successful completion" (OSD, 2018), our case study participants stated that ambiguity and disagreements between stakeholders still exist.

Communicating Within the Government Team

As discussed earlier, the government team that executes an OT includes the AO, requirements holder (e.g., AOR and potentially the PM), legal counsel, policy personnel, and all personnel supporting these roles. Early and frequent communication among members of the team is imperative to efficiently and effectively execute an OT. The AO and requirements holder need to work together on all aspects of OT execution. This includes, for instance, defining and refining the problem and strategy, determining whether an OT is the appropriate award instrument, deciding how to publicize and solicit sources, and developing the OT agreement. Additionally, many AOs we spoke with during our case studies indicated that requirements holders who are new to OTs may need to be coached on thinking outside of the FAR-based environment.

Integrating legal and policy personnel from the very beginning is also important. For example, legal counsel can help determine how to use OT reference material, such as the DoD *OT Guide* and previous OT agreements, and collect relevant information during market research and the solicitation process. Involving legal and policy personnel early is also a proactive measure, as it may mitigate the likelihood of potential disagreements at later stages that can delay execution and result in rework. Together, the government team needs to balance the potential value of the information it derives from the OT project with the cost, schedule, and legal risks. Working with legal and policy personnel early in the OT process should help to facilitate this shift in thinking and lead to smoother legal and policy reviews of decisions and documentation.

Two additional themes extracted from our case studies also have relevance to government team communication. First, as mentioned previously, many of our case studies experienced some turnover in government team personnel. Since there is no standard method for executing OTs, those personnel entering an OT process mid-execution may require additional communication efforts to get acclimated to the problem and previous plans made by the government team. Second, OT execution may involve government personnel from more than one office (e.g., when the requirements holder uses a contracting office from a different organization). In these cases, the interest in the OT project broadens, requiring approvals up more than one chain of command. Similar to the early involvement of legal and policy personnel, early communication between those offices can help to mitigate the likelihood of potential disagreements at later stages.

Communicating with Industry

A primary benefit of OTs is the unrestricted communication the government can have with potential offerers and awardees at all stages of the OT process.[5] As a result of this, our case study

---

[5] Note that the FAR contains strict rules on the types of interactions the government can have with potential offerers during the entire solicitation process to ensure all contractors can compete on a level playing field. For example, offerer clarifications about solicitations must be formally submitted to the government, and answers to those questions accessible to all potential offerers. As another example, after the open call for solicitations has ended, communication between the government and offerers is only permissible if executed in a manner that treats all offerers fairly.

participants told us that offerers have a better understanding of what the government wants, and so the government gets a better product. During the presolicitation phase, engagement between the requirements holder and industry (e.g., through market intelligence, presolicitation activities) can help to develop a better statement of need. During the solicitation phase, the more open communication reduces the time and effort an offerer spends trying to understand the solicitation compared with FAR-based scenarios and helps industry better prioritize their efforts. During the negotiation and award phase, a collaborative, iterative relationship with the offerer will allow for the government to achieve the full potential of what OTs can provide. Open communication at this stage also reduces the time and effort of all stakeholders because verbal negotiations can take place prior to formal contracting activities, which require legal and policy review.[6] This reduces the number of written iterations that may be necessary. The many benefits afforded through open communication are what attracts nontraditional sources to OT projects.

Because of the unique OT environment, early and frequent communication with industry is not only beneficial but also necessary. Nontraditional contractors are new to working with government; traditional contractors are new to working with OTs. AOs have to take on an educational role with contractors, guiding them on subjects such as IP and working with the government's acquisition environment.

Communicating with Other OT Users

As discussed above, many participants of our case studies stated that seminars and trainings, while important, were insufficient for learning how to execute OTs. They pointed to two additional resources: (1) on-the-job training with a mentor and (2) networking with other more experienced AOs. Each OT is unique, making the strategy for execution different every time. Learning about the execution of past OTs and how to think critically as new scenarios arise were two of the most important lessons for the AOs in our case studies. These lessons cannot be gleaned from compliance-based training. Instead, AOs in our case studies reached out to other OT users in their network. These users ranged from more experienced AOs within their directorate or command to those in other services. This was especially helpful to AOs who were executing their first OT or encountering a unique situation. Leveraging the OT experience of others is essential to effective OT execution.

Communicating with Higher-Level Stakeholders

A number of our case studies involved high-level stakeholders, such as the Office of the Secretary of Defense (OSD), Congress, and Air Force senior leaders. The reasons behind their

---

[6] At the same time, the flexibilities of an OT often add far more negotiations than would be needed in a FAR contract, which can increase the time of the negotiation process. While this aspect may be more time-consuming, our case study participants suggested that such additional negotiations were time well spent—the flexibility in negotiations attracts nontraditional contractors and can enable a better prototype.

involvement varied. Some of the OTs were awarding large dollar value projects and/or high-interest programs (e.g., NSSL), which attracted high-level stakeholders. Other OT projects sought out funding or other support through well-connected senior leaders. In some cases, high-level stakeholders acted as the overall impetus for the OT, explicitly directing the government team to use an OT for a specific problem. Whether high-level stakeholders were sought out or attracted to the OT project, communicating with them was a critical factor in ensuring a smooth OT execution.

While the involvement of high-level stakeholders is not unique to OTs (e.g., Congress routinely gets involved in high-interest defense acquisitions), the nature of the OT environment can create challenges when these stakeholders bring their own interests to bear. For example, ensuring that the OT complied with applicable statutes and policy was often a chief concern of high-level stakeholders. To address the concern, the necessary approvals to move forward with the OT project increased and, in some cases, required rework and delay. Sometimes these high-level interests actually helped to prioritize the OT project, accelerating the process. In all cases, early and frequent communications with these stakeholders were helpful in anticipating and proactively mitigating issues emerging from disagreements surrounding the OT.

As the government team formulates the problem, develops a strategy, and identifies funding for the OT project, anticipating high-level interests can improve smooth execution. Early and frequent communication with these stakeholders is likely to be the most important means of ensuring their interests are considered.

4. Observations About Developing the Problem and OT Strategy

In this chapter we use information and lessons resulting from the analysis of our case study interviews and supplemental documentation to discuss considerations that a government team overseeing an OT prototype program needs to bear in mind as it defines the problem its project faces and determines the strategy to answer the problem. As shown in Figure 3.1, these are the first two phases of the OT life cycle.

Defining the Problem

The first step in OT execution is defining the government's need. While this problem definition should occur in advance of deciding whether to use an OT, participants in many of our case studies suggested that it is essential for effective OT use. Many lessons we gathered from OT users for this step are applicable more broadly to prototype efforts. As was introduced in Chapter 2, an OT is not the only way the government can develop a prototype.

Statement of Need

The DoD user community often identifies mission needs statements, which are further developed into requirements and sometimes detailed specifications that are then applied to acquisition programs. However, OTs are exempt from the Joint Capabilities Integration and Development System process (DAU, n.d.), and at least partially as a result, a number of our OT project case studies stayed at the level of a *statement of need*. The distinction between "need" and "requirement" here is important. As in all acquisitions, performers under OTs are more likely to develop innovative solutions if the development problem is framed more on "what" you want to achieve than "how" you want to achieve it. If possible, requirements holders should be thinking about the goals or general outcomes of the effort, not the technical solution. Unlike the formal and specific requirements that are often produced during a traditional requirements determination process (e.g., key system attributes, detailed contractual specifications), the flexibility of OTs allows for the need to be written in nonprescriptive terms. This is crucial to finding innovative and diverse solutions and allows for offerers to be creative when developing proposed solutions. It also should provide for a more streamlined solicitation and award process because fewer iterations on the SOW may be needed.

A statement of need should be developed with some care. If it is too general, it is difficult to assess the level of success in an acquisition. By definition, development activities yield outputs that can be measured. This needs statement typically takes on the form of an outcome and depends on what the development community does and how users ultimately employ the outputs in new developments and integrate them with other well-defined systems. It is likely to lead to better outcomes for the user if it can be clearly characterized by metrics that translate the engineering outputs that developers control into operational mission outcomes that users value. The OT cases we examined had great difficulty creating statements of need that define the connection between engineering systems and operational outcomes well enough to assess clearly the success of OT projects.

By definition, a statement of need is user-oriented. But that should not lead OT practitioners to conclude that it is only relevant to acquisition programs driven by new operational requirements. An OT could very well be justified by awareness of a new technology that can meet user needs more cost-effectively than existing technologies. Indeed, one of our case studies included an innovative engine manufacturing process that was meant to significantly decrease engine cost. A statement of need remains relevant in such a "technology-push" OT. The value and success of the OT project still needs to be assessed in terms of how the technology that it develops affects the operational performance of its ultimate users. A new technology should not seek to justify itself in terms of engineering advances. Rather, advocates of using an OT to develop a new technology should show as clearly as possible how any engineering advance affects factors relevant to the user, whether they can best be stated in terms of, for example, improved military performance, reliability, or life-cycle total cost or ownership.

Revisiting the Problem Definition and Strategy Throughout the OT Life Cycle

It is worth noting that in almost all of our cases, the initial problem definition went through multiple iterations until an OT had finally been awarded. We found the process of defining the problem to be much more fluid in our OT cases than in FAR-based acquisitions we reviewed. The flexibility inherent in OTs allows for such evolution and iteration as information is gathered during market research and source solicitation and as negotiations take place prior to award. Unlike acquisition under a traditional development contract, our case studies also suggest that this flexibility allows (and in many cases requires) the prototype concept and design in an acquisition to evolve as information is gathered during execution of the contract following award.

Developing the Strategy

The next step in OT execution is to develop the strategy for soliciting sources and awarding the OT. Some (but not all) of our case studies developed a formal *acquisition strategy*. This is not required for OTs, but many participants stated that developing at least an informal conceptual strategy for the end-to-end approach improved outcomes throughout the process. This end-to-end approach may take the form of consecutive phases in which each step provides knowledge that can be used to set up the next step. For example, the RPS and Tetra OT projects each gathered information that could inform a second phase of prototype OTs. The Air and Space Operations Center (AOC) Cloud OT project used a sole-source follow-production OT (Project Kessel Run).

The end-to-end perspective could include phases for prototypes, production, and support. For example, an OT project can be used to collect data to inform a FAR-based acquisition. These multistep approaches may be especially helpful with complex developments since they allow for government and industry to evolve their solutions, adapt to market changes, and react to technical changes.

The government team can also begin at this time to consider whether the use of a cost-sharing approach is appropriate.[1] Cost-sharing not only reduces the obligations of government but also invests industry in the outcomes of the prototype. It helps align expectations with industry, improves outcomes (by focusing negotiations), and allows the market to work toward a more commercial solution.[2] While cost-sharing is required for traditional sources under 10 U.S.C 2371b, it can, and in some instances is, being used with nontraditional sources. Typically, these situations include those in which there is robust commercial development and/or venture capital funding that the government can leverage. Cost-sharing may also be used to shape the direction of research (but not fully set the agenda). However, it can also complicate that direction when defense and civil needs are not aligned.

We observed in our case studies a number of considerations and activities taking place to inform the overall OT strategy. The remaining sections in this chapter describe these subphases in more detail.

Identifying Available Funding

With the problem properly defined, the requirements holder next must identify funding. This can at times be more difficult for OTs than standard transactions because often the flexibility that OTs allow complicates their institutional alignment with existing plans or programs.[3] Given our sample of case studies, their response to new threats, and their innovative nature, funding for many of our OT cases was facilitated by a high-ranking, well-connected senior leader who took an interest in the effort. These leaders were either approached by requirements holders or were the impetus for the OT project in the first place. Many times they found funding to pursue the

---

[1] It is important to note that cost-sharing in OTs includes costs considered to be "in-kind." In the case of an OT, in-kind contributions are the value of noncash contributions (i.e., property or services) provided by either the government or contractor. As discussed in Appendix B, one of our case studies used in-kind contributions of resources to play the role of cost shares to meet the conditions set in 10 U.S.C. 2371b for conducting an OT with a traditional source.

[2] As further explained in OUSD(A&S) (2018), "This resource sharing requirement is intended to highlight the dual use focus of this authority and show commitment on the part of the performing team to pursue and/or commercialize the technology in the future."

[3] Since most OT projects to date and in our case study sample were not programs of record, they did not typically operate at the level necessary to be a separate program element in the Program Objective Memorandum. Thus far, OTs have typically been funded by other means as part of more flexible sources of funds. For further discussion on the types of funds that can be used for OTs, we direct readers to the "Identifying Available Funding" section of the DoD *OT Guide* (OUSD[A&S], 2018).

effort by combing their network. In summary, identifying funding for an OT project often requires entrepreneurial activities and networking.

Funding for an OT project can emerge from multiple government sources. But, as previously noted, each additional source brings its own interests to bear on the OT project. Further, requirements holders must consider the color of money and whether it matches the need. Independent research and development funds may be applied to OT projects, but we heard that operation and maintenance funds may not always be appropriate. Asking the comptroller to review whether the funding is appropriate for the OT project early in the process may prevent problems at a later date. One source of funds used in one of our case studies was the Small Business Innovation Research (SBIR) program. Using these funds provided the OT project with a stable, predictable source from year to year, which is not always the case for OT projects, but note that SBIR funds come in relatively small amounts.

Exploratory Market Research

For many OT projects, the market is undetermined. Nontraditional sources are, by definition, new to government endeavors and may therefore be completely unknown. In many of our case studies, exploring the market occurred in parallel or soon after the problem was defined. In some cases, the decision to use an OT was supported when initial market research revealed that only nontraditional sources could provide the necessary capability to solve the problem. In addition to helping shape the problem, exploratory market research can increase the awareness of nontraditional sources about government opportunities and help them understand how to participate. It has the additional benefit of potentially yielding broader information about capabilities beyond the defense industrial base that, in principle, the government can use more broadly to enhance its innovation and portfolio management work.

The value of market research to an OT project may be proportional to the government team's effort. Case study participants from two of our OT cases explained that more informed, detailed capability Requests for Information (RFIs) are more likely to encourage better offerers to participate and commit resources by giving them confidence that an acquisition is serious and could lead to an official program of record. In turn, more serious offerer participation in market research can yield better information on how to structure an OT. At the same time, the Air Force should consider that such outreach requires time and effort on the part of nontraditional sources. Our case studies suggest that the more calendar time it takes to complete a transaction with the government, the less likely nontraditional sources are to participate. In this way, increased industry outreach can also discourage nontraditional participation. The Air Force may need to find the appropriate balance to achieve optimal levels of participation.

Given the flexibility inherent with OTs, exploratory market research can take on many forms. For example, public events in relevant technology communities can provide information relevant to designing an OT and reveal potentially desirable offerers that an Air Force office has not done business with in the past. Other nontraditional methods that may be beneficial to understanding the market include social media, personal and professional networks, and professional associations' media. RFIs issued through Federal Business Opportunities (a.k.a. FedBizOpps or FBO)[4]—a government web-based portal that allows vendors to review federal procurement opportunities over $25,000—also can publicize potential OT projects; however, case study participants explained that most nontraditional sources do not monitor FBO. DIU takes a proactive approach by assigning dedicated personnel to engage in informal networking with the communities of technology companies around its three office locations (Silicon Valley; Austin, Tex.; and Boston, Mass.). Using these forms of outreach grows increasingly important as the OT becomes less similar to traditional FAR-based acquisitions.

Choosing Whether to Use an OT Mechanism

Usually after some initial exploratory market research has taken place, the requirements holder, supported by contracting office personnel, should begin to explore whether the use of an OT is both appropriate and the most effective award instrument. Similar to the problem definition stage of the OT process, this process is iterative. As funding is identified, the statement of need for the OT project changes, and as the market is better understood, the government team should revisit whether an OT is the best approach.

OT Goals

When deciding whether to use an OT mechanism (or any other award instrument), government personnel should first consider the goals of the effort. Table 4.1 provides a summary of appropriate reasons for using OTs as well as their potential secondary benefits. The latter should not be confused with appropriate rationale for the use of OTs. The appropriateness of specific rationale for OT use follows both from the congressional intent of expanding and permanently codifying the authority[5] as well as our case study participants' understanding of that intent.

As such, appropriate reasons for use include the ability to find nontraditional sources and/or innovative solutions to a government problem. OTs are one way to achieve these goals because they allow the government increased flexibility to solicit sources and award agreements in a way that attracts nontraditional sources and leverages commercial capabilities. Our case studies suggest that even when these goals align, government personnel still need to determine whether OTs are the most effective mechanism for the problem. There are other means of achieving these

---

[4] In the first quarter of FY 2020, FBO.gov will be decommissioned and transitioned to beta.SAM.gov.

[5] This intent includes making OTs "attractive to firms and organizations that do not usually participate in government contracting due to the typical overhead burden and 'one size fits all' rules . . . [and] support[ing] Department of Defense efforts to access new source[s] of technical innovation, such as Silicon Valley startup companies and small commercial firms" (U.S. House of Representatives, 2015).

goals using mechanisms such as the SBIR program and BAAs, among others.[6] Some case study participants told us that using an OT simply for the sake of using an OT is unlikely to yield the best outcomes for the Air Force.

Table 4.1 also lists a number of secondary benefits to the use of OTs that should not be confused with appropriate reasoning for their use, according to our case study participants. For example, OTs can allow for flexibility to avoid much of the FAR that is not relevant to a specific acquisition. That is a necessary but not sufficient basis for preferring an OT to an alternative approach. This secondary benefit, along with improving the government's ability to enter into cost-sharing agreements (e.g., because contractors are not required to use certified pricing data), can help to attract nontraditional sources.

Table 4.1. Appropriate Reasons for Using OTs Versus Potential Secondary Benefits

Appropriate Reasons	Potential Secondary Benefits
<ul><li>To attract and access nontraditional sources</li><li>To find innovative prototype solutions to problems</li><li>To flexibly design a solicitation and agreement to fit a unique problem or source</li><li>To leverage and influence commercial development</li><li>To most effectively address the problem</li></ul>	<ul><li>Eliminating unnecessary sections of the FAR/DFARS</li><li>Improving the government's ability to cost-share with industry</li><li>Decreased acquisition cycle time (but an OT is only sometimes faster)</li><li>Reduced exposure to industry protests</li><li>To allow for potential follow-on production using an OT</li></ul>

A prominent misconception that arose in our case studies is that improving speed is an appropriate reason to pursue an OT. However, we heard varying views in our case studies of whether OTs are faster than FAR-based mechanisms. Some types of prototype efforts may be awarded more quickly when using an OT, but not all. Many times, the calendar time saved during the solicitation and award process is equal to or less than the additional time that is needed to (1) define the problem properly and (2) perform market research on a market that is completely new to the government. As explained in later sections, the calendar time may be even longer if the OT involves government and contractor personnel who are new to OTs or if the OT agreement is developed without the use of a template. AOs in our case studies often had to act as gatekeepers, ensuring that requirements holders were pursuing an OT for appropriate reasons and not just because they wanted to move quickly.

---

[6] Sometime the most effective way to use OTs is when they are used in coordination with other mechanisms, such as partnership intermediary agreements, BAAs, and elements of the standard SBIR program and technology accelerators. Government personnel should also consider options for coordination.

Aligning with the OT Statute: Is the Effort a Prototype?

In addition to considering the goals of the effort, the other primary consideration when deciding whether to use OTs was whether the defined problem aligned with the OT statute as being a prototype. Government personnel need to work together early when defining the effort to decide whether it meets the prototype definition. If not, they may need to consider other award instruments.

The OSD provided definitions for a "prototype project" and "successful completion" in 2018,[7] yet our case study participants stated that ambiguity and disagreements between stakeholders still exist. Determining whether the effort is a prototype may be simpler when a physical system is being envisioned, but a few of our cases that involved processes or information technology struggled with coming to agreement. Government personnel told us that there is now general agreement that both business processes—such as prototyping a technology accelerator process—and physical processes—such as prototyping a new manufacturing process of an existing product—are considered to comply with OT statute. One AO provided three questions that government personnel should ask when trying to determine if the effort meets the definition for an OT prototype.

- What are we actually prototyping?
- What will contractors demonstrate?
- What does success look like?

That is, government personnel should be able to clearly (1) define the system, process, or concept being defined; (2) identify the steps contractors will take to develop and test the prototype; and (3) state the measures that will be used to evaluate the prototype.

Determining compliance with the OT statute may be complicated by high-level stakeholder interests. In a few of our case studies, senior leaders who either funded the effort or had vested interest in it advocated against the use of an OT because they questioned whether the effort could in fact be considered a "prototype." Additional deliberations were necessary to come to agreement. Communicating with these stakeholders early on in the process and recasting the scope of work, if necessary, can help to anticipate and mitigate such complications.

Contracting Decisions

Considering Options for Contracting Offices and Other Facilitators

The requirements holder of an OT project and the contracting office executing the OT project need not be a part of the same directorate, command, or even service. Three of our case studies included Air Force–funded OT projects that were awarded by Army contracting offices.[8]

---

[7] See Chapter 2 for a full definition.

[8] The Technology Accelerator case study was originally awarded out of ACC Aberdeen but was transferred to AFRL after award.

What this boils down to is this: OT requirements holders have choices of the contracting office they use to execute the OT. This choice depends on a number of factors, including whether the requirements holder organization has a warranted AO. In at least two of our case studies, the requirements holder organization did not yet have a warranted AO, which was at least a partial impetus for seeking out external contracting services.

Our case studies suggest a number of other considerations that may be useful when deciding where to contract the OT.

- *Experience:* AOs and contracting offices that have more experience using OTs are likely to be more effective and efficient. There is a learning curve to effectively using OTs. If the OT is complex, a contracting office with significant OT experience is more likely to have encountered a number of the unique or challenging aspects that might occur during execution. However, if the OT involves a simpler endeavor, there is an argument to be made to use internal contracting options so that local OT experience can be built.
- *Relationships:* Results from our case studies suggest that a good working relationship among all government personnel involved in an OT is essential to its effective use. An advantage of using internal contracting services is that these relationships may have already been cultivated. This may result in the OT potentially moving faster and obtaining a better product overall for the government.
- *Coordination:* As more stakeholders are involved in executing an OT, the amount of coordination and iterations to obtain approvals increases. This can be further exacerbated when those at the working level must coordinate with more senior personnel to receive approvals. Coordination adds calendar time and may increase the amount of rework necessary. When requirements holders and contracting personnel work for the same organization, this can reduce the number of stakeholders involved and approvals necessary to execute the OT.
- *Expertise/Domain Applicability:* While contracting personnel do not need to have technical knowledge about a specific technology area or industry to execute an OT process, OT projects from similar domains often share characteristics that allow for efficiencies in execution. That is, OT project solicitations and agreements are tailored to the unique problem and industry. Once this tailoring occurs with one OT, lessons can be applied to future OTs with similar characteristics. Often, the locus of specific domain knowledge such as this would reside with a requirements holder's internal contracting office. One reason, however, to use an external contracting office or facilitator would be if they have a unique ability to access sources specific to that domain. DIU, for example, specializes in accessing sources in Silicon Valley. OT consortia also can provide access to sources from specific domains. The decision of whether to use an OT consortium will be discussed in subsequent sections of this chapter.
- *Cost:* There is the possibility that the use of external contracting offices, especially those outside of the Air Force, could mean that the requirements holder will have to pay a processing fee to cover the costs of that external contracting office. This did not occur with any of our case study OTs awarded through the Army Contracting Command, but we learned that the possibility does exist and is worth considering when choosing a contracting office.

Considering Facilitating Organizations

In addition, two of our case studies involved facilitating DoD or Air Force organizations—DIU and AFWERX—that performed some of the OT executing functions.[9] For these two case studies, these intermediary organizations provided OT support to the requirements holder and contracting office but are still internal to the DoD.[10] These intermediaries offered OT development expertise such as more extensive market research, a structured solicitation process (DIU's commercial solutions opening [CSO] for example),[11] and some execution support. In this case, these intermediaries offered a compromise between using an external and internal contracting office as they combined the benefits of control through an internal OT development and management effort with a greater level of OT expertise available through an external organization.

Deciding Whether to Sole-Source the OT project

OTs provide the government team with flexibility to make a sole-source award to a contractor with virtually no required documented justifications for the sole-source decision.[12] This may be attractive in unique circumstances where market research identifies that only one firm that can provide the OT project system or service. Out of our seven case studies, only one was a sole-source OT project. In that case, the awardee was a small company that had previously held a subcontract to supply an engine for two large prime contractors developing a cruise missile for the Air Force.[13] The OT project involved designing a manufacturing process for that engine. The past experience of the awardee with this specific engine meant that no other sources could perform the effort. In that case, the AO formally documented this justification, even though no mandate existed.

Using a sole-source OT in these situations has the advantage of reducing the time and effort that the government team incurs to get the OT project awarded. However, choosing this route also requires that any follow-on OT project production activities related to the prototype must be competitively sourced. Another one of our case studies initially considered a sole-source OT award for similar reasons—the contractor to perform the effort had already been identified

---

[9] In addition, the Consortium for Command, Control, and Communications in Cyberspace (C5) at the Army's ACC-NJ, which awarded the TAP Lab and OBAC project, uses the Combat Capabilities Development Command to facilitate some OT executing functions.

[10] In contrast, consortium managers, who also provide support for OT projects awarded under a consortia, are external to the DoD and act as contractors.

[11] Note that with the passage of Section 879 of the NDAA for FY2017 and the DFARS OUSD(A&S) Class Deviation, 2018-00016, the use of CSOs to solicit for OT awards may no longer be allowable.

[12] OT statute requires that competition be used to the "maximum extent practicable," but neither the OT statute nor DoD's *OT Guide* state that documented justifications are required (10 U.S.C. 2371b; OUSD[A&S], 2018).

[13] The awardee, however, was in fact, a nontraditional defense contractor since their contract for this engine was entered into with the two prime suppliers. Prior to this OT, the awardee had never held a direct contract with the government.

and had a unique advantage over other firms because of unique past experience. In the end, the government team chose to compete the OT project but to develop a more targeted solicitation. This provided two benefits. First, the solicitation ensured that the team did not miss any possible sources. Second, the government could use a sole-source production OT. To enable an efficient solicitation process, the government team developed a targeted solicitation that would reduce the likelihood of receiving and having to evaluate noncompetitive offers.

Deciding Whether to Use an OT Consortium

Among the many flexibilities that OTs provide, the ability to use an existing OT consortium to solicit for and award an OT project is one worth exploring. Two of our seven OT cases were awarded under an OT consortium and case study participants from at least one other case expressed that they might have used this option if some of the consortia that are available now had been established at an earlier time.

Our case studies suggest that government teams should bear in mind a number of considerations in deciding whether to execute an OT project through an existing consortium and choosing which consortium to use. These considerations include the following items.

- *Access to sources*: To effectively execute an OT, the government must be able to publicize the needs or requirements to the most relevant sources. If an existing OT consortium aligns well with the problem, using it may be the most efficient means for the government to publicize its needs. If there is a mismatch, the requirements holder may not reach the most relevant contractors, in which case the development of a standalone OT may be more appropriate.
- *Flow-down terms and conditions*: When a project is awarded through an OT consortium, most of the terms and conditions for that award are predetermined through flow-down provisions in the base agreement with the consortium manager. The government team should review these flow-down clauses prior to choosing to use a particular OT consortium. While some clauses, such as IP, may still be negotiated, others are relatively fixed. In one of our case studies, the requirements holder used an OT consortium that awarded only firm-fixed-price agreements for a project that may have been best awarded under a cost-plus agreement, leading to additional costs to the government.
- *Potential for efficiencies*: An OT consortium can present the government with a number of efficiencies that may not be possible using a standalone OT. Given their predetermined terms and conditions and structured processes, a consortium may be able to execute an OT project in less calendar time than when using a standalone OT. Many track the average time it takes to get an OT project on contract, which may be helpful when deciding whether to use one. As discussed further in the next section, consortium managers also can support a number of government functions in relation to OTs, such as screening proposals, supporting the development of SOWs, and performing administrative functions. While many of these functions would be performed by the AO of a standalone OT, they may also reduce some burden for requirements holders.
- *Cost*: Consortia may charge two separate fees that will affect requirements holder costs: (1) a fee the requirements holder pays to the government consortium program manager for using their service and (2) an administrative rate paid to the consortium manager that

is deducted from a project award. The two OT consortia projects in our case studies only assessed fees for the latter. Administrative fees for consortium managers vary across OT consortia.

- *Consortium contracting office*: Requirements holders should also consider which contracting office is executing the OT consortium. Many of the contracting office considerations outlined in the previous section apply to the choice of using a consortium.

Deciding Whether to Establish a New OT Consortium

If demand in a particular domain is high enough, there may be reasons to consider establishing a new OT consortium instead of having a contracting office execute a number of standalone OTs. Our case studies illuminated a number of considerations the Air Force should weigh when deciding whether to establish a new OT consortium. These considerations are summarized in Table 4.2 as advantages and disadvantages to contemplate. Overall, when deciding whether to establish a new consortium, the Air Force should weigh the benefits gained from efficiencies (including better access to appropriate sources and more effective OT execution support) against the cost of hiring a consortium manager and the up-front effort to establish the consortium.

Table 4.2. Advantages and Disadvantages to Establishing an OT Consortium

Advantages	Disadvantages
• Consortium managers reduce burden on government (e.g., screening proposals, administration of OT projects, educating nontraditional contractors how to work in OT environment) • OT consortia allow for better access to a targeted industry or domain, and in turn, may have better insight into commercial technology development within that industry or domain • OT consortia create efficiencies by establishing predefined flow-down terms and conditions for industry consortium members, reducing effort and calendar time during OT project award negotiation • Consortium managers can provide surge capacity when demand is high • As more awards are made under OT consortia, consortium manager fees should be reduced • Experienced consortium managers can provide support to contracting offices new to OTs	• Additional government costs associated with consortium manager administrative and management fees • Large time and effort investment for initial establishment • Consortium manager adds an intermediary stakeholder, potentially reducing government control • Decisions made up front will affect all OT projects awarded; challenges in amending base agreement once consortium members have signed on • Some consortium managers are for-profit firms, which could create conflicts of interest in their execution of tasks • Could be seen by contractors as creating a "pay-to-play" environment (i.e., in which contractors have to "pay" consortium fees to "play" in the consortium's competition), especially if new consortium domain is partially redundant to an existing one • May require higher-authority approvals to establish a large ceiling amount (i.e., total dollar value that can be obligated under a consortium) • Can potentially inappropriately favor members of a consortium as sources of services to OTs relative to sources that are not members

As described above, the use of predetermined terms and conditions likely will save government time and effort during negotiations. However, establishing those clauses requires considerable up-front work. When project awards do require negotiations, the consortium manager can act as a liaison with the contracting office, which should reduce that office's burden. Consortium managers can provide assistance to industry by helping answer solicitations and educating sources on how to navigate the OT environment. They can have surge capacity when demand is high because they are able both to hire new staff more quickly than government and to shift personnel between multiple consortia they oversee. These services come at a cost to both government and industry. For the former, a percentage of funds for every project awarded is paid to the consortium manager. These fees also pay for consortium managers to access, maintain, and grow their industry membership. Since most nontraditional sources do not monitor FBO, the use of a consortium can be an efficient means of advertising OT project opportunities.

When the Air Force decides to establish a new OT consortium, our case studies present a number of lessons for consideration.

- Creating a consortium around a specific technology area or industry allows the model/ process to be tailored to that area. There is no one-size-fits-all consortium model; the creation of broadly focused consortia may introduce unnecessary problems.
- The consortium manager administration fee is based on both fixed and variable costs. As more OT awards are made under a consortium, its ceiling is increased, and/or other consortium OTs are awarded to an existing consortium, it may be possible to negotiate the fee downward.
- There is an entire industry that performs consortium management services in both FAR and non-FAR-based settings. Some of its members are for-profit while others are nonprofit. Many use different management models. Many have been managing consortia for decades. Government personnel establishing an OT consortium need to grasp an understanding of this market.
- In certain mission areas, where large-value projects are often awarded, establishing an OT consortium with a "low" ceiling (i.e., one that only requires local approval levels) may create roadblocks when demand exceeds the initial ceiling. In these cases, getting authorization for a higher ceiling from the start may be beneficial.

5. Observations About Soliciting, Awarding, and Executing OTs

Once exploratory market research is complete, the government team can begin to solicit sources, select a recipient, negotiate and award the agreement, and finally execute and close out the agreement. These phases, depicted in Figure 3.1, constitute the post-market-research elements of the OT life cycle. In this chapter we discuss considerations—informed by analysis of our case study interviews and supplemental documentation—that OT users should bear in mind as they navigate these latter aspects of the life cycle.

Source Solicitation and Selection

Solicitations

With a better understanding of the market, the government team may undertake an interim selection process that can also inform the final solicitation. A majority of our case studies used a two-step down-selection approach to finding sources. We learned that it is also a standard approach used for OT projects being solicited under OT consortia. The first step, which often can be a Request for White Paper, casts a broad net and asks for a limited amount of information. Requests are usually for short papers (anywhere from two to ten pages) and often come with turnaround times of two weeks to a month. The information received allows the government team to down-select offerers and, thereby, to better inform the scope of the subsequent solicitation. The second step, sometimes in the form of a Request for Prototype Project (RPP), asks for more detailed information on capabilities of greatest interest. These often allow for lengthier responses, are sometimes tailored to specific offerers based on their response during the first step, and provide more calendar time to respond. Given the flexibility inherent in OTs, engagement between the government team and offerers can occur throughout this two-step process.

Our case study participants explained that the two-step approach can benefit both the government and contractors. During the initial step, it allows offerers to participate and the government to evaluate offerers while expending minimal resources. Responses from the first step can also inform the government—before it incurs large sunk costs—if it missed the mark when it developed the statement of need. Finally, it opens communication channels between the government team and potential second-step offerers so that they can better tailor their RPPs to their unique situation, which can reduce rework and evaluation time during the selection and negotiation process.

Not all OT project solicitations need to use a two-step approach. A few of our case studies were able to gather enough information during exploratory market research such that the first step was unnecessary. Regardless of whether they used a one- or two-step solicitation approach, our case studies suggest a number of strategies for solicitation development that should improve the effective use of OTs.

- Write the statement of need using nonprescriptive language in nonmilitary terms. Nonprescriptive language will allow for broader, more innovative solutions to be proposed. This may attract nontraditional sources by giving them a greater ability to benefit from their proprietary creativity. Using nonmilitary terms also may improve finding these sources as the solicitation will be more accessible to those without military backgrounds.
- Post solicitations in a number of places. In addition to the nontraditional publication methods (e.g., social media, professional associations' media) mentioned in the previous sections, the government team should consider whether an additional posting on FBO is appropriate. Since traditional contractors can still participate in OT projects in cost-sharing scenarios or with significant nontraditional participation, an FBO posting would help to ensure their awareness. When soliciting an OT project through a consortium, posting the solicitation on FBO increases the transparency of consortia proceedings. It may also provide a mechanism for consortium member recruitment. The FBO posting can direct industry to the consortium manager for further information.
- Provide a SOW template for offerers to include with their response. Since many nontraditional sources are new to working with government, providing a template for solicitation responses may improve the quality of their responses, reducing the need for clarification and rework. A number of our OT project cases provided a template for the SOW, which we learned was standard practice for OT projects solicited under consortia. The draft SOW can then become the foundation for the technical terms in the agreement. This can reduce the calendar time needed to reach agreement with an offerer on the agreement.

Selecting an Offerer

After proposals have been received, the government team can begin to evaluate each and select a recipient. Similar to FAR-based source selections, the requirements holder organizes a technical evaluation of each proposal and submits the results to the contracting office and office of general counsel for review. One important consideration for the government team during this process unique to OTs is whether the potential recipient meets the conditions for award as prescribed in 10 USC 2371b. If not using a cost-share OT, awards can generally only be made to nontraditional sources or traditional sources with significant nontraditional participation. There are a number of characteristics to consider, but we learned that some firms bidding for OT projects are able to achieve nontraditional status either because (1) they are a subsidiary of a traditional contractor but are organized in such a way that they still qualify or (2) they have "significant" nontraditional participation.

Unlike most FAR-based acquisitions, the selection of an OT awardee includes some additional flexibilities worthy of consideration. First, a number of our OT cases operated the selection process similar to that of a BAA. That is, the government team made multiple awards, each with a very different award amount and SOW. Using this approach allows the government team to test multiple prototypes to see which works best to solve the problem. There is greater ability to operate a selection in this way when the solicitation uses a nonprescriptive statement of need, as discussed in the previous section. In most of our cases, the government team had planned and announced in the solicitation that it would make multiple awards. However, in one of our cases only one award was originally planned. Two offers were received for which the government team saw value in pursuing. Because a well-networked senior leader had interest in the OT project, the government team was able to find additional funding to make two awards.

One additional flexibility with OTs not afforded by most FAR-based transactions is the ability to begin negotiations with an offerer prior to making a final award selection. In one of our case studies, this approach was used to ensure that the government team could come to an agreement with the leading offerer before making a final selection. We also learned that such a strategy can allow for the government to use other offers as leverage during negotiations with the leading offerer, similar to negotiations that take place in the commercial sector.

Agreement Development

Approach to Developing the Base Agreement

One key feature of OTs that distinguishes them from FAR-based transactions is the flexibility available when developing an OT agreement. Rather than being required to conform with the specific terms and conditions in the FAR (e.g., cost-accounting requirements), AOs may tailor and negotiate the terms and conditions to the specific details of the prototype and awardee. This can be especially advantageous in a number of OT-relevant circumstances, such as when working with a nontraditional contractor that is accustomed to commercial practices, when standard FAR IP clauses would overly restrict a contractor's efforts, or for unique business arrangements like consortia.

For each of our case studies, we discussed with participants the approach used to develop the OT agreement and supplemented those discussions by reviewing the agreement. Our research suggests that AOs use a spectrum of approaches to develop the initial OT agreement (i.e., prior to negotiation with the prospective awardee), as depicted in Figure 5.1. At one end of the spectrum, a number of AOs apply a *template approach*. These AOs seek out a past OT agreement that is relevant to their particular problem (e.g., similar technology, cost-share scenario, consortia, etc.), use that agreement almost as a template, and tailor it as necessary to their unique problem. At the other end of the spectrum are AOs who apply a *synthesis approach*. They begin with a blank page and fill it in using multiple resources, such as example agreements, the DoD *OT Guide*, or basic contract law. They choose relevant clauses from each and tailor the overall agreement to the unique problem.

Each approach has advantages and disadvantages. AOs who lean toward the template approach may do so because it can be accomplished quickly and more simply. If the agreement

Figure 5.1. Spectrum of Approaches for OT Agreement Development

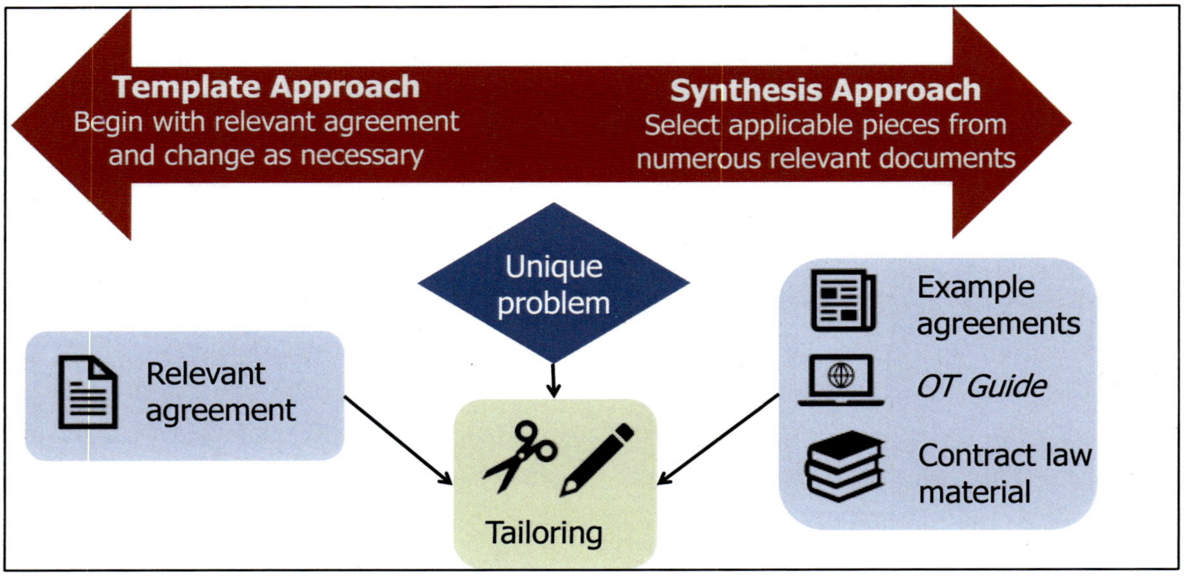

they are using as a template was awarded from the same office, they know that it has already passed legal and policy review, potentially reducing the number of iterations necessary. AOs gravitating toward the synthesis approach may be better able to leverage the full flexibility allowed with OTs. It provides them with an agreement that is better tailored to the situation, which may result in a better product. Since the approach follows an opt-in situation, in which clauses are chosen and added to a blank slate, it is more likely to reduce the number of unnecessary clauses in the agreement.

Regardless of the approach chosen, the government team may reference a long list of helpful resources when developing an OT agreement. These include past OT agreements, the DoD *OT Guide*, SBIR language, contract law material, past assistance agreements, cooperative agreements, and traditional contracts. The FAR/DFARs can also be used as references when developing certain relevant clauses for an OT. For example, data rights clauses may need to be developed based on the unique situation. They will be more useful when the government team can (1) resist translating them into a checklist and (2) strike an appropriate balance between using historical terms and conditions because they have proven to be successful in the past and tailor new terms and conditions to the unique circumstances of the OT. Successful past experience designing and executing OTs can improve the ability to strike this balance appropriately. Early legal and policy participation can also help ensure these resources are used properly.

Agreement Terms and Negotiations

When developing the terms and conditions for an OT agreement, our case study participants echoed one common theme: start early. Early negotiations with the offerer can help to tailor how the Air Force shares risks with the offerer. Given the flexibility of OTs, much of the negotiation can take place in a verbal forum, possibly reducing (but not eliminating) the amount of legal and policy review at the very early stages.

Our case studies suggest that the Air Force can consider presenting the offerer with a list of nonnegotiable clauses at the outset to reduce iterations on certain clauses. This method has especially been successful for OT consortia, as it often requires potential industry members to agree to nonnegotiable clauses before they can become consortium members. Clauses based on FAR language may also be acceptable to many nontraditional sources. Even though we learned that the reason a number of nontraditional sources are more attracted to OT projects is because they do not need to comply with the FAR, we also heard that many reject the FAR in theory but less so in practice. That is, some nontraditional sources will accept an OT even though it requires elements of the FAR that these sources seek to avoid by using an OT. One strategy used by a few AOs we spoke with is to begin with federal clauses required in any agreement and then negotiate with each offerer the minimum set of terms and conditions necessary to achieve the government's objectives.

One primary deviation from this logic is with terms and conditions related to data rights, indemnification, and liability. Many of our cases required negotiations in these areas. Liability appeared to be more of an issue for small companies out of concern for their long-term viability because a large-value OT project could make up a large percentage of their work. Data rights for intellectual property, however, was an area of negotiation for most of our cases, and we learned that trying to secure government control of IP can significantly lengthen the calendar time required to negotiate an OT. A major reason for these negotiations was not because of disagreement but because of inexperience with data rights for IP in the OT environment. Many AOs reported that they needed to educate both contractors and requirements holders about data rights. They often had to play the role of an impartial mediator. Contractors were sometimes too willing to provide unlimited government rights while requirements holders sometimes were too restrictive with these rights when they may not have needed them. One AO stated that IP negotiations were smoother when they could help contractors identify (1) their existing proprietary IP, patents, and inventions and (2) the rights they are delivering to the government. This helped to clarify what rights were negotiable.

Another area where negotiations are common is within the SOW and payment milestones. In many of our case studies, the offerers submitted a proposed SOW with their proposal. Once an offerer is selected, our case studies suggest a number of considerations for the government team to account for in negotiating a SOW. First, the final SOW must still meet the definition of a "prototype" according to the OSD policy. Negotiations may result in changes that bring this compliance into question. Second, the SOW should offer a balance on its level of specificity. It needs to be written with enough specificity that the contractor has confidence that it will be paid for the work executed. At the same time, care should be taken to provide the contractor with flexibility as the OT project unfolds and more information is gathered. One of our cases had to delay execution of the OT to amend the agreement; the government team thought it might have been avoidable if the SOW had been properly written in the first place.

Finally, milestones and associated payments also require up-front thought. Written well, milestones can manage industry cash flow and motivate deliverables. Negotiation can address the cost to the provider and value to the government of each funding milestone. OT completion metrics and payment milestones should have the ability to discriminate success from failure. At the same time, "successful completion" does not have to translate to a working prototype. OTs can be written as best-effort agreements and not be task oriented. This is attractive to industry because it reduces their risk. The government team in one of our case studies found that zero-dollar milestones or milestones for inconsequential activities (e.g., holding a meeting) were ineffective. Similar to many FAR-based transactions, properly defining the cash flow is important to nontraditional contractors, as they often do not have enough working capital to begin the effort without an up-front payment or to meet payroll during execution. Payment terms may also require additional effort for government teams new to OTs, as a few of our cases reported having to iterate with the Defense Contract Management Agency (DCMA) and Defense Contract Administration Services to ensure existing payment infrastructure could accommodate the unique situations that arise with OTs.

Executing and Closing Out the Agreement

Once the OT agreement has been awarded, the performance can begin. While a number of our case studies did encounter some problems during OT performance, most could have been corrected in earlier stages of the OT life cycle (e.g., changing language in the original OT agreement). For these cases, we include those lessons and perspectives in the section of the report that aligns with the specific life-cycle phase in which these problems could be prevented. However, regardless of the specific execution problem, we heard that OTs allow for flexibilities that reduce the calendar time and effort necessary to address problems—flexibilities not available in FAR-based transactions. Adjustments to the OT agreement, whether they be in funding, in cost-share percentages, or in the SOW, are relatively simple and do not require higher-level approvals or large amounts of paperwork. Milestones and payments associated with them can be changed dynamically to address impending problems. This flexibility is especially important in the OT prototype environment as additional information gained during early stages of performance may shift the needs for later stages. The important message we heard from case study participants was to remain flexible during execution and to continue to engage in close communication with the contractor.

Another important point raised by some during our case study discussions was that the role of the PM (and more broadly, the government) changes for OTs. With a traditional FAR-type transaction, the PM's role is more rigid in making certain the contractor is delivering. However, with an OT, the role of the PM can shift to being more of a partner and helping the contractor.

Given that OTs are generally best-effort agreements and not oriented toward specific technical deliverables, the government can more easily, for example, offer technical assistance if the contractor runs into problems compared with traditional FAR approaches. Establishing such a partnership should improve the product of the OT project. This consideration stresses the importance of frequent communication between the government and contractor.

Conclusion

The preceding three chapters provide a structured set of considerations that proved useful for our case study participants. Since each OT is unique, these considerations should be thought of in a similar light. They should provide the government team with some options but are not an exhaustive list. At all times during the OT life cycle, critical thinking and business acumen need to be exercised for whether and how to use OTs.

6. Policy Considerations in Employing OTs: Goals, Culture, Environment

In the three previous chapters, we discussed observations and lessons for implementing OTs based on our case studies and discussions with various stakeholders. Still using these discussions as our foundation, we shift gears in this chapter and focus on observations and issues that are more relevant to senior policymakers in the Air Force. In particular, we discuss whether the Air Force is achieving the goals associated with OTs, the cultural impediments to successfully implementing OTs at scale, and needed changes in environment to better leverage OTs for prototype development. Many of the issues and suggested changes were brought to our attention explicitly by our case study participants. We further inform the discussion of these issues and changes by incorporating considerations offered by relevant literature and our own research experience in defense acquisition.

Is the Air Force Using OTs Effectively?

OTs are one of a set of tools that DoD seeks to use to address a concern that near-peer competitors have set an accelerated pace of innovation that threatens the U.S.'s ability to sustain its technological dominance. The more OTs can do to help DoD respond quickly to near-peer innovation, the better.

With Congress having expanded OT authority in recent years and making the authority permanent in the FY 2016 NDAA, the DoD has now renewed momentum to pursue three complementary objectives to create conditions that can help yield this counteradversary outcome.

1. Expand the use of prototypes to (a) explore risky alternatives, quickly and at a potentially lower cost, and (b) increase the longer-term likelihood of finding high-performance solutions to demonstrated operational needs.
2. Attract innovative sources of goods and services that have not traditionally done business with DoD. In particular, seek firms that are using cutting-edge technologies to enter or even disrupt competitive commercial markets.
3. Help expand its relationships with traditional and nontraditional sources. Removing the FAR requirements in OTs potentially reduces the administrative cost of working with DoD borne by nontraditional sources. The loosened rules under OTs also create more opportunities for nontraditional firms to innovate in ways that would benefit DoD.

Each of these objectives complements the others. Flexibility and the chance to add real value increase the likelihood that nontraditional firms can make money working for DoD without compromising their commercial opportunities. Nontraditional firms are likely to have ideas, developed to exploit competitive commercial markets, that DoD's traditional sources do not have. An incremental approach can manage the overall risk associated with trying risky ideas in each increment, which will redound to the mutual benefit of the government and its sources.

OT Lessons from Our Seven Air Force OT Project Cases

Are Air Force–sponsored OT projects being used effectively? It is too early in the Air Force's pursuit of such OTs to get a definitive answer. But the seven case studies we examined in depth offer insights about whether Air Force–sponsored OT projects can be used effectively. In this section, we provide a summary of our case study findings about the Air Force's use of OTs, which was expanded on in the previous three more detailed chapters.

Our cases tell us that they can and that many are performing well against the objectives set out above. But much depends on how individual OTs are executed. The Air Force is continuing to learn how best to use OTs as it accumulates experience. Maximizing the future potential of OTs in the Air Force will depend in part on the Air Force's ability to continue learning and take as much advantage as it can of the potential that OTs offer. Many of the practices that help the Air Force benefit from OTs (e.g., early and frequent communication, nonprescriptive statement of need) could help it benefit from other approaches to acquisition—even application of standard FAR-based acquisition. But these practices take on special importance in the flexible environment that OTs make available.

All of our cases address an OT for a prototype, which means that they have demonstrated the Air Force's ability to conduct development activities to limit risks prior to production. Only some of our cases have displayed iterative risk reduction through successive prototypes. And it is too early to assess follow-ons that may have resulted from these prototypes or to ask how the prototypes shaped these follow-ons. Even in the absence of any follow-on, a prototype may have served its purpose well by generating the information needed at the time.

Evidence from our case studies suggests that OTs have allowed the Air Force to create agreements that more closely resemble true commercial contracts than a contracting officer (CO) can typically compose by using standard FAR clauses. Nontraditional firms need not understand the FAR to understand these agreements. And under these agreements, opportunities exist for richer communication between the Air Force and a contractor than standard FAR procedures typically allow. This has permitted greater reliance on informal, sometimes oral, agreements and less reliance on formal documentation during interim stages of OT development. It has allowed quicker agreement on mutual terms. Negotiation is not always just about prices and deliverables; it is also about establishing language, relationships, and terms that leave both sides comfortable within their familiar professional contexts. It has allowed greater flexibility in the definition of the requirement for the final deliverable, increasing the likelihood that the deliverable will match operator needs when it is ready for delivery. It has allowed more subtle and flexible treatment of IP, allowing the Air Force to get the information, rights, and fairness it needs without endangering the future commercial value of the IP to a nontraditional source.

In principle, an OT allows each agreement to be closely tailored to the circumstances at hand. In practice, Air Force OTs have often been built on prior agreements.[1] These reflect a balance among the purely commercial terms a contractor might want, the requirements of federal law even when the FAR is not in play, and the experience of the AO with what language has worked well in the past. Getting this balance right depends on continuous learning and on the AO's empowerment and self-confidence working in a fluid environment. The success of getting an OT that attracts nontraditional sources while not requiring too much time and effort to tailor depends heavily on the AO's skill and their ability to work with legal, finance, requirements, and other specialists to get the balance right quickly.

The balance is easier to get right if AOs have market research that tells them what sources look most attractive to the Air Force and what terms and conditions are likely to attract such sources.[2] AOs must know where to look for the right sources. This may require accessing unfamiliar networks and various forms of professional and social media. Once AOs have found the community of sources, they must know how to elicit the information they need. The information is not just about contracts. It potentially covers the ecosystems in which companies in this community assess capital and do business with their customers. Meantime, the AOs' approach to market research, in itself, can inform high-quality nontraditional sources about how good the government might be as a customer or partner. Success is more likely if AOs have a long-term relationship with legal, finance, requirements, and other relevant stakeholders who can explain the information they would value and if AOs can translate the priorities of the Air Force into terms that nontraditional—and in particular, nonmilitary—technical communities can understand.

Perhaps because part of DoD's current mantra is to go faster in acquisitions, we learned during our case studies that some Air Force personnel, particularly requirements holders, have come to OTs with the expectation that they can speed up the pre- and postsolicitation award process. Our cases show that this process can go faster in some cases. And all else being equal, nontraditional contractors prefer quicker execution of agreements. But that is not the intended use of an OT. Tailoring an agreement typically takes time to achieve its full mutual benefit. In the long run, achieving that benefit is likely to be what attracts nontraditional sources to the Air Force. As the Air Force learns more about commercial contracting standards, it should be able to craft mutually attractive agreements more quickly. But it is still learning, and the models of contracts that commercial firms use will continue to evolve over time.

---

[1] Consortia often develop terms and agreements that are essentially nonnegotiable; each new task can be tailored, but it must accept the core set of terms. Similarly, as Air Force organizations have gained experience, they have often built comparable core terms and conditions of their own. Or they have borrowed such terms and conditions from other organizations with more experience.

[2] Of course, market research might well start in an acquisition before a decision is made to pursue an OT. Presumably information from market research should inform that decision. Our discussion here refers to market research specifically relevant to designing an OT.

The flexibility found between the Air Force and a contractor in an OT often carries over into the bureaucratic relationships used to define requirements and pursue acquisition within the Air Force. Because OTs are typically not bound as closely to standard requirements and funding processes as traditional FAR acquisitions are, they tend to rely more heavily than FAR-based acquisitions on informal networks to define requirements and obtain funding and then to assess the final deliverable. This flexibility can affect schedule and final performance in ways that are technically outside the purview of an OT agreement itself and the AO's authority.

The flexibility inherent in OTs can create the potential for Air Force benefits, but only if government and industry pursue mutual gains (i.e., both sides benefit as a result of the agreement). Traditional FAR-based contracting obviously seeks mutual gains, but typically uses a very structured approach to achieve this objective. Because OTs are typically less structured, the government and contractors rely more directly on mutual trust and reputation to work toward these mutual gains. The Air Force has found that some government personnel and traditional industry sources steeped in traditional FAR-based contracting can have difficulty adjusting to such a flexible environment. This inability to shift to a more flexible contracting environment makes achieving a beneficial arrangement for both sides problematic. For example, some Air Force contracting personnel view themselves as stewards of the Air Force's money, and as such, cannot conceive of a contractor making more than a typical government rate of profit, even if doing so would yield exceptional outcomes for the Air Force. Such win-win outcomes (higher profits for higher Air Force gains) are standard among best-in-class commercial firm agreements.

By the same token, our cases also revealed that small nontraditional firms unfamiliar with the government had difficulty understanding how to protect their IP in a government setting. An AO had to assume the role of an advocate for a contractor to protect the contractor from its own ignorance of the law. Air Force OTs face special challenges finding nontraditional sources that are well prepared to work productively with their government counterparts.

As OT projects get larger in value, the number of stakeholders and level of stakeholder interest increases. Both complicate the AO's ability to work flexibly with the contractor and with legal, finance, requirements, and other stakeholders in the Air Force because there are more and stronger stakeholder perspectives to accommodate. Higher-level interest has the potential to discourage risk-taking, especially if it comes from outside the Air Force. The result: opportunities to exploit OT flexibilities and mutual benefits become muted as the OT project gets larger or exhibits a higher profile. Changing the willingness to use OTs to bear more risk within the Air Force is already a challenge. Standing up to risk-averse oversight from outside the Air Force increases the challenge.

The majority of observations above are not unique to OTs. DoD has conducted prototypes to reduce risk for decades without OTs. FAR can accommodate many elements of standard commercial contracting. Skilled COs can tailor FAR-based transactions to find flexibilities with standard clauses. Market research to understand commercial markets can be valuable in a traditional FAR setting. Traditional requirements and acquisition processes are encouraged by the DoDI 5000.02 to be adjusted to allow more flexibility. FAR contracting arrangements can accommodate useful negotiations that lead to win-win outcomes for all parties. Our point here is that OTs' flexibilities enable greater use of these sorts of actions compared with the FAR, and that increased flexibility is what attracts advocates. Thus, in the absence of actions like these, the advertised benefits of using an OT are much less likely to accrue. Without them, an OT is not special.

Defining Success

The discussion above of our case studies naturally raises the questions of how the Air Force should define success when using OTs to pursue prototypes.[3] This section discusses this question and was informed by discussions with OT users.

As noted above, DoD views OTs as one tool available to help the United States respond quickly to accelerated innovation by potential adversaries. It likely will not be possible to generate defensible, empirical measures of this kind of effect for any particular OT. But we can use the proximate objectives discussed above to define more easily defensible measures of success.[4] Once measures of success have been defined, the natural next step is to develop the infrastructure to reliably collect, track, and analyze such measures. This too will likely present challenges for the Air Force to implement and, therefore, is worth its further consideration.

Expanding the Use of Prototypes

Getting a good measure of this requires the definition of a counterfactual: Would a prototype have occurred if an OT had not been available? The answer to this is likely to depend on subjective judgments of how much a counterfactual prototype, conducted without an OT, would have cost; how long it would have taken to set up and execute; and whether the counterfactual prototype would have generated information as good as the prototype conducted with an OT. It would also depend on subjective judgments of how creative the team conducting a counterfactual prototype could have been in the absence of an OT and whether the OT was as creative as it could have been given the flexibility an OT provided.

Eliciting reliable judgments of such a counterfactual scenario could require significant effort, but it would be worthwhile on a sample of completed OT-based prototypes for two interrelated reasons: (1) such judgments could generate a valuable, periodic program evaluation of the

---

[3] An important nuance to this discussion is that "success" for a prototype is different from "success" for an OT. OT "success" may be partially measured by whether the prototype project was successful, but this alone would not answer the question of whether using an OT to develop the prototype was the most effective mechanism or whether congressional intent for expanding and codifying OT authority is being achieved.

[4] Translating objectives into measures that the Air Force could use to improve the performance of its OT program will be challenging. Any specific metric must be understood in context, given the way that managers will use it to drive performance. Our case studies gave us only limited insight into how to do this in the OT program; further work is warranted. For an overview of the challenges involved that might provide guidance to future Air Force actions, see Baldwin, Camm, and Moore (2000).

Air Force's OT program; and (2) perhaps more important, the judgments required in such an evaluation should help the Air Force identify better ways to pursue prototypes with or without OTs.

Accessing a Nontraditional Source

As long as a nontraditional source is well defined, this metric of success is easy to measure. A more interesting question would ask if a nontraditional source was a good one to access. The Air Force could address this in at least two ways.

- Did the nontraditional source come from a technological community that the Air Force considers desirable for some reason? For example, is the source engaged in a technology outside the traditional defense industrial base that offers promise to DoD? Is the source well subscribed by venture capitalists? Do the principals in the source have a reputation for developing disruptive innovation? None of these questions is likely to yield a complete picture. Multiple questions could yield valuable information.
- Did the nontraditional source generate an innovation that a traditional source likely would not have developed? This is a harder question to answer because (a) the immediate product of a prototype is not typically a good predictor of value over the longer term and (b) comparing a prototype with one that a traditional source might have conducted requires developing a counterfactual, with all the challenges mentioned above. Waiting longer might allow for a more complete assessment of the value of information generated by a prototype, but waiting is hard and probably inadvisable when innovation is accelerating in near-peers adversaries. Moreover, as more time passes, more and more factors enter that could "explain" technological advances that a prototype might facilitate. These difficulties are inherent in any effort to value an innovation in light of the costs of generating it.

Loosening Traditional Rules

Measuring this may also require a counterfactual: What rules would the Air Force have applied to a prototype in the absence of an OT? In principle, given the task of designing a prototype, a textbook case could be constructed to explain a nominal approach in the absence of an OT. This approach could then be compared with the approach actually taken using an OT. Presumably, the greater the difference between the approaches, the greater the possibility that the availability of OT authority would have supported a loosening of many traditional rules.

A program evaluation could stop there. But it would be desirable to take the next step and ask this question: What was the effect of using flexibilities only available in an OT? In a specific completed OT, did these flexibilities allow a prototype to move faster, proceed at lower costs, and generate more valuable information? Or did it compromise important discipline and allow negative outcomes that many traditional rules were specifically designed to deter or prevent? Similar to the questions posed above, these are not easy to answer. They require subjective judgment—although perhaps less than those above do. But it would probably be worth the cost of answering these questions for a sample of OT-based prototypes after they have been completed.

The answers would help the Air Force (1) periodically evaluate its OT program and (2) develop lessons learned on what rules to loosen and how to loosen them.

Adjusting the Culture of Air Force Acquisition

Over time, DoD has evolved a complex approach to managing the development and procurement of new systems. The approach seeks technological breakthroughs that have historically given the United States the most technologically capable systems in the world. However, DoD has policies centered around managing risk. Oversight of these policies has evolved to create incentives that drive conservative behavior that tends to give more attention to following the most conservative (risk-averse) options rather than rewarding prudent risk-taking and substantive outcomes. These incentives have created an ethic that emphasizes stewardship of public monies and needs to protect taxpayers from contractors' intent on profiting from the government. This risk-averse, conservative options approach is increasingly dated as the pace of threats increases.

OTs are one tool created to help DoD pursue and gain technological breakthroughs in a more nimble, commercial-like manner. The more DoD is helped in this realm by OTs, the more its culture becomes like that of its nimble, commercial high-technology counterparts. It cannot be completely commercial because government still has requirements to "solicit and assess potential [OT] solutions . . . [using] a fair and transparent process" (OUSD[A&S], 2018), but some prudent moves to more commercial-like transactions and associated risk-taking appear to be helpful (as exhibited by the cases studied in this report).

Willingness to Take Risks

Commercial firms view risk differently than DoD does. If a company's product is late to market in a competitive industry, some other company will likely dominate that market. This risk is baked into any risk calculus that a company uses to assess its decisions to develop and produce a new product and to manage it as technology and market needs change. A company accepts some additional risk on meeting individual performance, cost, and schedule milestones if doing so reduces the risk of being late to market. In particular, an established company understands the desirability of making many small "failures" if they teach it how to identify the product that will ultimately help it dominate a market. A company sustains latitude to explore many options, often by testing them to failure, to find the one that can lead to a final success. A company rewards its personnel who take the risks of failing to generate the information that the company will ultimately benefit from. The incremental and sequential prototyping that OTs can facilitate, of course, are natural ways to pursue this approach.

Private firms potentially have the benefit of being allowed to invest their own capital in one year and to gain from that investment years later when it finally leads to their success. They also potentially have access to bankers, venture capitalists, and other partners who can provide capital and benefit with a company when it succeeds years later.

It is hard to promote such risk-taking in any large bureaucracy. DoD cannot afford performance risks that put missions or lives at risk. Also, DoD has no readily available capital and cannot borrow for high return-on-investment options. It must go to Congress each year for funding. Moreover, Congress often expects near perfection in acquisition rather than rewarding prudent risk-taking.

A number of our case study participants expressed some hesitancy with the risk-taking necessary to effectively use OTs. We were told that while they saw senior leaders promoting taking such risks, the cultural shift had not flowed down to their own organization. Experience with such congressional oversight has built up in DoD and encouraged DoD acquisition personnel, from senior leaders on down, to avoid the risk of failing. Ironically, it can be easier to associate a small failure with an individual than to say who is responsible for failing to introduce a major new capability before a near-peer adversary does. DoD needs to give more attention to managing risk in the broader context of this near-peer global competition. It must give its personnel the latitude to fail and learn as they explore the way forward.

Spreading this approach across DoD and the Air Force in particular will require extensive change in the way that supervisors manage their subordinates, in how personnel systems assess and promote government personnel over their careers, in how inspectors general and auditors assess personal and organizational performance, in how senior officials react to congressional oversight, and in how COs choose and manage the contractors that create and produce most of the technology that DoD relies on to perform its mission. In each of these venues, DoD has evolved standard risk management practices and policies that sustain what participants have all come to accept as normal and nominal. But what is accepted as normal and nominal must change in each of these venues for DoD's approach to managing risk to change. Until that changes, the gains that the Air Force can realize from OTs may be limited.

Managing Process or Outcomes

The FAR and DFARS can be thought of as intricate, standardized options books that acquisition personnel can use to work through any acquisition-related task. If they follow one of those options, they are compliant. Experienced personnel know that these standardized options can be used creatively to achieve remarkable results. But the mainstream guidance for using these options seeks to protect less experienced personnel from making errors that could easily threaten outcomes. Most of the formal classroom and online training available to DoD acquisition personnel teaches how to be successful in compliance with these mainstream guidelines (e.g., DAU, 2019). Moreover, risk aversion leads acquisition personnel to appreciate the sense of safety that compliance with mainstream guidance can give them.

An alternative approach mentioned by a few of our case study participants would place less attention on compliance and more on the realized outcomes. This is easier to do in the commercial world where an individual or office's rewards can be tied to the outcomes that the individual or office realizes. Performance is also typically easier to measure in the commercial world where sales, profits, and market share dominate. And there is more latitude in linking compensation systems to performance measures associated with an individual or office. The training and incentives that personnel in the commercial world receive gives little attention to compliance (except those required by law) and much more attention to problem-solving in specific and dynamic situations that employees are likely to encounter across their careers.

Much of OTs' flexibility aims to emphasize realized outcomes and puts less weight on compliance-oriented rules and checklists. DoD can do three things to help government personnel exploit the flexibility that OTs allow to get better outcomes.

- Change the way its personnel perceive and manage risk, as discussed above. The personnel should associate risks more with outcomes and less with compliance or the elimination of all risks. They also should be incentivized to include the risks associated with near-peer global competitors (i.e., threats to the warfighter) in their risk calculus.
- Ensure that formal acquisition training and mentoring give less relative attention to compliance and more to problem-solving. DoD can benefit from doing this throughout all acquisition training, but it is especially important for training and guidance that prepares personnel to design and manage OTs.
- Accumulate experience that helps acquisition personnel understand how to take advantage of flexibility in new settings. Doing this likely will help personnel develop and manage better OTs. Workforce shaping should keep this in mind as it prepares acquisition personnel to become AOs and, along the way, to perform contributing roles that support an AO. We give this more attention in the material on workforce shaping.

Building Partnerships to Share Mutual Gains

Two qualitatively different relationships exist between buyers and sellers in markets for goods and services. One tends to be *transactional and short term*. Each party seeks a simple exchange that requires little information sharing or investment in the relationship. Each understands that the other will gain little from cooperating or sharing any gains realized from the exchange. Such transactions occur most often in the commercial world in simple settings where products are well defined and transactions trade money for equivalently valued goods and services in a supply-and-demand context; they occur in a "zero-sum" world. The other is more *strategic and longer term*. The buyer devotes effort to understanding the seller's capabilities; the seller devotes effort to understanding the buyer's needs. The two invest time in finding opportunities that will benefit both—opportunities likely to be tailored to the relationship and superior to what either party could expect from an impersonal, short-term transaction. They can also potentially seek partnerships for mutual gains without concern for fairness and openness to other potential buyers and sellers.

The FAR in its simplest form presupposes the former kind of relationship. Full and open competition will yield a deal to be executed at arm's length in a setting where neither party fully trusts the other, and each seeks to receive any mutual gain that the two might share. The FAR provides tightly controlled rules to manage such an exchange. Also, while the FAR recognizes and enables some long-term relationships (e.g., down-selection to a single prime contractor at a certain point or even federally funded R&D centers with long-term relationships), it seeks to ensure that the companies involved continue to perform acceptably and seek future competitions when possible to incentivize performance and seek equitable prices over time. Moreover, the governmental concerns for fairness and transparency as well as social concerns such as support for small, disadvantaged, minority, or veteran companies force broader considerations beyond private pursuit of a partnership with a single company.

Analysis of our case studies suggests that OTs can enable a more commercial relationship of the latter strategic and longer-term approach. It allows for information to be closely exchanged, for the terms of the agreement to be tailored, for the two parties to be flexible about deliverables as they learn from one another, and for both parties to have a stake in the relationship such that they will reap mutual benefits. Experienced personnel can use the FAR to nurture this latter form of relationship; an OT facilitates more direct flexibility to pursue such a relationship when appropriate.

Different kinds of buyers and sellers frequent the markets that support these two kinds of relationships (e.g., Moore et al., 2002; Camm, 2005). In particular, sellers seeking a strategic opportunity spend little time on the former type of relationship because it allows little room for the buyer and seller to create something new that will benefit them both. When a buyer frames a solicitation in terms that apply simple FAR language, sellers seeking relationships that are more compatible with their IP, pricing, or other equities learn to stay away. They tell those who will listen that "they don't do business with the government." They even fear that working with the government could degrade their reputation in the broader marketplace with other companies seeking strategic opportunities. Indeed, we heard many of these perspectives from our case study participants. The result is that a traditional FAR setting, particularly for services, tends to draw sellers with more transactional interests.

This natural separation in the market has consequences for those seeking to use OTs in the Air Force. Our case studies suggest that nontraditional sources the Air Force might want to attract with an OT are sensitive to the signals that an Air Force buyer sends when it frames a solicitation and executes an agreement. The Air Force's behavior creates a reputation that sophisticated sellers will be aware of. To benefit most from an OT, the Air Force needs to seek the more strategically oriented seller. A few of our case study participants stressed the importance in the OT environment of the Air Force maintaining a reputation as a similarly strategic buyer seeking opportunities to create gains that both parties can share. That reputation will make it easier for the Air Force to interest potential sellers in participating in the kind of market research likely to result in a mutually successful OT. We heard that how the Air Force conducts such market research can send signals to potential sellers about how serious the Air Force is about mutual gain and about the levels of effort potential sellers should put into participating in Air Force activities. The reputation also makes it easier for a seller to assure the other buyers it works with remain a strategic partner whom they can continue to trust.

The importance of the Air Force's reputation among the firms it wants to attract to an OT points to an important "externality" that senior Air Force acquisition personnel should monitor as the Air Force conducts OTs. How an Air Force–related OT that is conducted today treats its nontraditional counterparts has consequences for other Air Force–related OTs in the future— consequences that go beyond or "are external to" today's OT. That means that one outcome of an OT conducted today is the way that today's OT affects the reputation of the Air Force as a whole as a partner in OTs. Senior Air Force leaders should include this outcome among those it monitors as it manages its enterprisewide portfolio of OTs.

OTs as a Means of Conveying Cultural Change in Air Force Acquisition

Any discussion of creating organizational change is difficult to operationalize. However, our observations about OT usage suggests that Air Force leaders are potentially doing so already by signaling their support of OTs generally. The acquisition workforce has responded to that signal by designing and executing OTs. These OTs may feature some elements reminiscent of FAR-based contracting (e.g., the template versus synthesis approach described in Figure 5.1), but on the balance are more agile and responsive to the acquisition and technology environment.

This signal and response mechanism may represent an organizational culture shift from what Wilson (1989) describes as a *procedural organization* that focuses on measuring process adherence to a *craft organization* that focuses on outcomes. Here, Air Force leaders' support of OTs serve as a tonal shift in the policy environment, as described by Wood and Waterman (1993). The acquisition workforce knows that OTs are new processes; they might also then think through their choices and decisions about any contract or agreement more deliberately instead of being weighed down by decades of organizational inertia and previous practices, even if those previous practices (i.e., FAR-based acquisition) could support more agile and responsive acquisition practices. Instead of requiring a cultural shift to fully take advantage of OTs, OTs may be a catalyst of the organizational change that is sought.

Building an Environment for Using Prototype-Based OTs More Effectively

A basic insight about effective agile management states that personnel feel more comfortable to take risks when they operate in a well-designed environment that prepares them to use the flexibility at their disposal and supports them in a predictable manner; rewards them for taking prudent risks and for managing risks objectively; and enables them to learn from (rather than punishing them for) failures. Put another way, successful agile behavior rarely arises spontaneously. It flourishes in an environment carefully prepared to facilitate agile behavior.[5]

---

[5] For a discussion of such preparation, see Kim et al. (2020).

The same insight applies to the effective use of flexibility in OTs.[6] This section discusses four elements of preparation, informed heavily by our case study discussions, that can help provide an environment designed to facilitate better use of prototype-based OTs.

- Shaping the workforce
- Sharing information to support continuous learning and improvement
- Managing consortia from an enterprisewide perspective
- Adjusting administrative tools to reflect flexibility of OT-based prototypes.

Shaping the Workforce

Success with OTs depends heavily on the reliable availability of a skilled, experienced workforce. Our research suggests that the Air Force will benefit from proactively shaping its acquisition, technical, and legal workforces to create and sustain a mix of skills and experience that will help it use OTs more effectively in the future.

Learning Through Training and Experience

The Air Force currently relies heavily on formal training—in the classroom and online—and accumulation of appropriate experience to build and sustain the skills that it needs in its acquisition workforce. DoD experience with OTs suggests that the Air Force can build on this approach to create and sustain the different skills it needs to create and manage effective OTs.

Current formal Air Force training focuses heavily on teaching how to comply with the FAR and DFARS. OTs can use materials from the FAR and DFARS, but OTs exist precisely to avoid taking a compliance approach to using these materials and would rather (1) use only the materials needed for a specific task and (2) work with a potential source to tailor the terms of an agreement to its and the government's mutual benefit.

Training today provides little support for such work. Our discussions with DARPA and the Army Contracting Command, New Jersey (ACC-NJ), for example, suggest that the most effective training uses case studies and role playing to lead students through decisions they will need to make to create and manage effective OTs. Ideally, the cases resemble a situation that the students will face as members of an integrated team in the near future. That means that the training includes all the functions—acquisition, technical, and legal—relevant to creating agreements for prototypes. For example, ACC-NJ offers a regular two-day course that teaches the basics of OTs on the first day and conducts a case study with role playing on a potential OT

---

[6] Of course, it applies to DoD acquisition as a whole as well. As we have noted, our focus here is on OTs. It is possible that the institutionalization of the approach we support here could yield an increasingly burdensome bureaucracy that could progressively limit the flexibility that the Air Force seeks with OTs. Some might argue that such a process has yielded the FAR-based culture that OTs seek to avoid. This concern should be taken seriously; awareness can offer the first defense against repeating this process. For a useful discussion of the broader effects of this process on productivity, see Olson (1982).

on the second day. Students identify the case to be used in each course. ACC-NJ believes that the training is more effective when all functional players participate (e.g., Anderson, 1999).

Such training can resemble mentoring in support of an upcoming task. Most individuals we interviewed felt that formal mentoring was the best way to train personnel on OTs. Currently, a great deal of the AO training is done through informal networking, in which a new AO reaches out to other AOs in their network to learn some OT basics, for example. Some more formal mentoring processes and perhaps the creation of a center of excellence as the focus for OT knowledge for the Air Force might be beneficial in the long run.

More broadly, everyone we interviewed with deep experience with OTs agreed that formal training alone cannot prepare the personnel in the functions that must work together for an OT to succeed. Rather, a person most effectively learns how to create and manage an OT by creating and managing an OT on the job. This approach imagines engaging new personnel who want to learn how to conduct OTs as they occupy junior roles where their supervisors can mentor them as they build their skills in a low-risk, well-monitored setting. The Air Force has relied on this approach to build the skills of its traditional acquisition personnel by giving them increasing responsibilities through their careers. But the Air Force must recognize that the skill sets it is building differ when it is preparing personnel to use the FAR and DFARS for inspiration rather than as a safety net that tightly constrains how the Air Force manages a contractor. Accumulating experience is important to all the skill sets relevant to building an OT but particularly to the skills that an effective AO needs.

Contracting Officer (CO) or Agreement Officer (AO)

Today, those who oversee the creation of OTs in DoD are actively seeking the personnel available with the most experience and, in particular, a rich enough understanding of both government and commercial acquisition to operate effectively outside the formal bounds of the FAR and DFARS. We heard during our case study discussions that this is leading to an unexpected discontinuity in shaping the acquisition workforce, including in the Air Force and, specifically, in Air Force contracting. A system of shaping that has built the skills over time needed by the CO for a complex acquisition is effectively being preempted to divert the people with these skills from a traditional career path to one that leads into OT management. It is too early to say how serious or widespread this change is, but it raises several issues for senior leaders.

First, how separate should the workforce responsible for OTs be from those responsible for more traditional Air Force acquisition? As noted above, an OT is only one tool that the Air Force can use to manage acquisition. Presumably, the CO for a program should be prepared to compare an OT with alternative approaches and apply an OT in circumstances where that is appropriate. This suggests that the CO should have the skills to conduct an OT, should have someone on their staff with those skills, or should have access to a center of excellence that can help them determine whether to use an OT. If the CO relies on a center of excellence, then they may need to be prepared to engage someone with the skills to conduct an OT and to manage that person.

The Air Force is just now encountering these decisions for the first time and has no preferred institutional way forward. If the Air Force expects its traditional acquisition, technical, and legal workforces to add OT skills to their own skills, they likely must give up other skills.[7] What skills are no longer needed? If the Air Force wants to build new elements of its acquisition, technical, and legal workforces with specialized OT skills, it needs to formalize what is effectively a new career field—or at least a marker on a person's personnel record that allows the Air Force to track their experience in OTs. And it needs to formalize the relationship between the traditional workforces and the new OT-oriented workforces.

If the Air Force forms specialist OT career fields, it must decide how to manage them. As elsewhere in DoD, the Air Force is currently diverting more experienced personnel from traditional responsibilities to OT responsibilities as needed. If OTs become a standard Air Force tool, the Air Force will face persistent competition for its most experienced personnel, who at some point could ask or be asked to move from a steady upward progression in traditional acquisition to a specialized career. Because this specialized career requires experienced personnel, it could come to be seen as an elite corps with special status. If that occurs, what will be the relationship between the CO and AO associated with a particular program? And will someone who has moved from the CO to the AO career track be willing to return to the CO career track to take jobs with higher levels of responsibility? The creativity that an AO needs can also be valuable in more traditional but complex acquisitions. A difficult balance must be achieved so that the entire contracting portfolio is executed effectively.

We heard questions of this kind in our interviews. The answers may sort themselves out as the Air Force gains more experience with OTs and understands how extensive their use will be. The answers may have better outcomes for the Air Force if senior leaders anticipate these questions and proactively assess alternative ways to respond to them. Here, we simply present these types of questions for the Air Force to anticipate and assess—we are not advocating for either path.

While we have framed these questions in terms of the relationship between COs and AOs, the same issues arise elsewhere in the acquisition, technical, and legal workforces that will continue to support OTs as long as the Air Force uses them.

The Air Force is also still working through how to issue an AO warrant. Each Air Force organization we talked to seemed to treat this a bit differently. Some AOs have warrants

---

[7] That is, the government adds skills to its employees by investing in their human capital. To add a skill, the government must increase its investment. It can do this by reducing current work requirements to give an employee the time to learn the new skill, reducing what it invests to accumulate other skills, or increase the total skill level of its personnel with a commensurate increase in compensation to retain employees with enhanced skills. Adding skills is not costless. It is likely that adding a skill reveals a change in priorities about what the government values—a change that is likely to reduce the priority that it places on other skills.

commensurate with their CO warrants (typically a dollar threshold). Others hold unlimited warrants. Still other AOs may hold a warrant for a specific project. Given the flexibility inherent in OTs, it is not clear that more uniformity in AO warrants is needed. However, this may be a barrier to the ability of AOs to move between organizations. This will likely resolve itself as the Air Force gains more experience with OTs. Again, the resolution may have better outcomes for the Air Force if senior leaders anticipate this and proactively assess alternative ways to address it.

Our interviews also identified risks associated with working on OTs and, in particular, with becoming an AO. A good AO needs to have a flexible, innovative understanding of the intent of the FAR and DFARS and be comfortable taking risk. But will an AO be rewarded for taking risks? Or could they face career consequences for taking OT-related risks that fail or run into problems? This raises the question addressed above about whether the Air Force can rise to the occasion of treating risk-taking appropriately. The question is aggravated by the fact that, as noted above, to date, the Air Force has not articulated any specific way to identify when an OT should be judged successful. Performance evaluations of AOs and others supporting OTs need to be reconsidered.

Stability in the Workforce

Turnover of government staff can be more problematic for OTs compared with standard contracts. Much of the activity in an OT is done through negotiation and discussion and is less reliant on formal documents or procedures. Hence, when a key individual, such as a PM or AOR, turns over, issues that have been previously decided or agreed on may need to be reopened. Documentation on the rationale for decisions may not exist, leading to the potential for additional negotiations and discussions. Also, there is a reeducation process for the new team member that might be required regarding the specific terms of the agreement. One can't expect a standard set of terms and conditions, for example. A number of our case studies experienced some kind of staff turnover, resulting in the issues discussed above. Our interviews emphasized the value of keeping the acquisition, technical, and legal personnel associated with an OT as stable as possible to allow them to work flexibly together as they created and then managed an OT. Poor stability tends to degrade the outcomes of an OT if they are measured in terms of performance, cost, or schedule. Because this degradation is beyond the control of any individual associated with the OT, the importance of stability to OT outcomes increases the risks to individuals when the Air Force judges their individual performance when associated with an OT.

Retention of AOs and Others with OT Experience

Our interviews indicated that demand is high for individuals with experience performing OTs. Different parts of DoD are seeking candidates from each other's talent as OT activity expands and each component gives increasing attention to servicing OTs motivated by requirements generated within its own component. Successful participation on OTs also gives personnel expanded visibility outside the government where the skills associated with an OT are highly prized and, according to our case study participants, compensation may often well exceed that in DoD. Like the rest of DoD, the Air Force faces an increasing challenge as its own demand to conduct OTs rises and therefore may have difficulty retaining the talent generated in the OTs it has conducted to date. The Air Force has faced similar challenges in retaining pilots and others with highly valued skills. Its experience working to retain those with valuable skills could be beneficial here as well.

Sharing Information to Support Continuous Learning and Improvement

The OT communities in DoD, and in the Air Force in particular, are currently engaged in a period of active innovation and learning, which is likely to continue as interest in OTs grows. Individual offices are developing their own approaches to OTs. Our interviews revealed significant differences from office to office. Some offices, like DARPA, DIU, and ACC-NJ, have accumulated enough experience that they have become de facto centers of excellence. In each case, a few now-experienced individuals have effectively taught themselves how to use OTs and shared what they had learned with colleagues as they joined in the continuing creation and management of additional OTs. Other offices pursuing OTs often consult informally with these centers. But personnel at each of these centers told us that their own best practices continue to evolve to such a degree that they are reluctant to document and share best practices, inasmuch as they expect only a subset of those practices will remain valid for any length of time. And they fear that any formal documentation could be interpreted as a template for building OTs that would defeat the central purpose of OTs—creating agreements anew in new settings. Each office encourages creativity and improvisation to realize as much benefit as possible from new OT opportunities as they arise.

This pattern of continuing innovation and diffusion of knowledge is familiar to private-sector and government organizations that have participated in the various versions of the quality movement that have arisen, morphed, and spread since the 1980s.[8] We observed in our case studies that information about OTs appears to spread mainly through ad hoc personal networks that individuals in separate offices use to share information. The result is that effective knowledge sharing is limited. The Air Force would probably benefit from promoting more proactive information sharing.[9] Information sharing could be facilitated both within the Air Force OT user community and between the Air Force and OT users in other services and DoD agencies. The broader defense acquisition community may be facing similar challenges with OTs. The Air Force may benefit from learning how those organizations manage and mitigate such challenges. Further still, these organizations may have encountered other challenges or

---

[8] These have included, for example, total quality management, Lean Six Sigma, and agile management and acquisition. See, for example, Camm (2003, 2005, 2006) and Kim et al. (2020).

[9] Since concluding our case study research, SAF/AQC and DoD have begun to explore strategies for information sharing such as providing a reference list of all warranted AOs. We did not learn of any case study participants having knowledge or access to this list yet.

scenarios the Air Force is not currently anticipating. This situational awareness may help the Air Force as it shapes the role OTs will play in the future.

Strategies for effective information sharing vary. However, experience from the quality movement suggests that sharing information on networks is likely to be more effective than sharing information on specific practices or policies. The practices and policies that work best in specific circumstances are likely to differ, particularly in the pursuit of OTs seeking to tailor agreements. Written descriptions of a practice in a current setting are less likely to be useful in a new setting than personal discussions with experienced individuals who can factor in nuances about how a current practice might behave in a new setting. And practices and policies thought to be best in class today will likely look dated in five years. Organizations have tended to prefer knowledge management strategies that share contact information among offices pursuing new practices rather than details on what new practices look like at these offices (e.g., Camm et al., 2001).

Managing Consortia from an Enterprise-Wide Perspective

Consortia have grown to provide marketing services on a wide range of topics potentially relevant to prototype-based OTs. They provide valuable administrative services to government agencies that prefer not to perform such services related to purchased services in-house. They provide other valuable services connected with coordinating partnerships among private firms seeking government opportunities and advising these firms on how to deal with the government. And for all participants, they offer natural marketplaces where buyers and sellers of related services can find other parties with common interests. In return for this value, government and private-sector participants pay a variety of fees that differ from one consortium to another. The exact nature of the fees appears to be evolving over time as consortia and their government and private-sector customers learn more about the services they provide.[10]

When government buyers decide to pursue OT-based services through consortia, the consortia effectively become gatekeepers for access to the DoD OT marketplace.[11] There can be good reasons to allow consortia to become gatekeepers. The organizations that manage consortia have accumulated many years of experience providing relevant services. This experience allows them to provide these services cost-effectively. As all participants in OTs learn how to use them in a more effective manner, consortia can potentially help focus and capture this learning as a part of

---

[10] Compare information on the cases of the Air Force Flexible Acquisition and Sustainment Tool and Army Rapid Response to Critical Systems Requirements Program in Camm, Blickstein, and Venzor (2004).

[11] When the government selects a consortium, it is in effect making a judgment about how it wants to maintain its access to relevant sources and choosing a consortium with the characteristics that best meet its needs. The selection is analogous to selecting a prime contractor that the government expects to assemble and manage subcontractors to produce what the government values. It is analogous to selecting a contractor that will then serve as the sole source for a series of tasks to be determined later, over the course of the contract. Governmental concerns about openness and fairness need to be fully addressed in the initial selection of any consortium.

their administrative support. But some of our case study participants raised the question of whether the services that consortia provide to participants, especially smaller, nontraditional sellers, are worth the fees that they charge. Others worry that multiple consortia are arising to service markets for similar services. Maintaining membership in multiple consortia can potentially limit full participation by smaller companies with a limited ability to tolerate overhead costs.

These concerns have raised questions about the degree to which DoD—and the Air Force in particular—should (1) rely on consortia rather than directly managing OTs with private firms and (2) regulate consortia to avoid redundancy among consortia trying to service the same areas of the market for access to government opportunities. These questions deserve the attention of the senior leadership. Based on the empirical information we have collected in this study, we cannot endorse any specific answers to these questions. But we can comment that, when some observers refer to multiple consortia serving the same market as being redundant, others see these consortia as competitors. General federal acquisition policy favors competitors in part because, if a competitor fails to provide services worth the cost of its fees, the companies it seeks to attract—including small businesses—are likely to look elsewhere if another competitor is available. In the presence of such competition, it may be easier for the government to decide which consortia to rely on for non–inherently governmental administrative support.

Adjusting Administrative Tools to Reflect Flexibility of OT-Based Prototypes

Our interviews identified a number of standard DoD acquisition tools that Air Force personnel have had difficulty applying in the flexible setting of creating and managing OTs. Today, these personnel are making do with a variety of creative work-arounds. As OTs become a more standard part of the DoD approach to acquisition, senior Air Force leaders may want to support adjusting these tools or developing new ones better suited to use in the management of OTs.

Contract Writing Software

At the time of our research, DoD's contract writing software, ConWrite, was not set up to allow AOs to write OTs that are tailored to their specific projects and to the flexibilities allowed with OTs.[12]

Payment systems

OTs often involve nontraditional defense contractors that do not have FAR-compliant cost and accounting systems. One government case study participant suggested that disbursement problems can be problematic for some contractors that do not have robust cash flows and require

---

[12] Since concluding our case study research, SAF/AQC has begun an initiative to introduce new contract-writing software to the Air Force acquisition community. Case study participants were not aware of this initiative, and it is yet unknown whether it will provide the needed flexibilities to align with OT execution.

timely payment to remain in business. Because of difficulties with DoD contracting and payment systems designed for more traditional procurements, participants from a few of our cases stated that they had to implement work-arounds or repeat reviews to expedite payments. Such disbursement problems may create government inefficiencies and could discourage nontraditional contractors from working with government in the future. A potential lesson is to review the contracting and payment systems and processes to ensure that they support the easy implementation of OTs contracts.[13]

Federal Procurement Data System-Next Generation (FPDS-NG) reporting

FPDS-NG is the standard common database that DoD uses to report and store data on contracting transactions. It collects data on standard contracts and OT agreements through separate processes. Those processes used to collect information for FPDS-NG on OT agreements are insufficient for collecting data unique to the OT environment. For instance, the data fields available for reporting in FPDS-NG are unable to capture awards for prototype projects under OT consortia. Discussions during our case study research and relevant literature (e.g., Schwarz and Peters, 2019) suggest that the resulting data on OTs are less reliable. They are also more dated (i.e., delayed in reporting) than data on standard contracts. Further, current data fields in FPDS-NG are unable to collect the data required to comply with the formal reporting requirements passed in draft legislation of the 2020 NDAA (Williams, 2019). As OTs become more common in DoD, full integration of data collection on OTs into FPDS-NG would provide more complete insight into the extent and uses of OTs.[14]

---

[13] Of course, any changes to payment systems and processes would need to be coordinated with existing organizational processes, which could present a number of its own challenges.

[14] Past the basic contracting metrics reported in FPDS-NG, further research is needed to identify OT performance metrics that could be collected and analyzed to further inform policy decisions. As discussed earlier in the chapter, defining OT success, and in turn, performance is not without its challenges. However, OT performance metrics are likely needed to fully understand OT implementation and possible improvements.

7. Conclusions

OTs for prototype projects are an award instrument that have seen a dramatic increase in use over the past few years. Throughout this document, we have described a number of observations and lessons concerning the use of OTs by the Air Force based on a series of focused case studies and discussions with the broader Air Force and DoD acquisition community. These observations and lessons are geared to both senior policymakers and practitioners. While our focus was to inform Air Force practitioner and policymaker decisionmaking, these lessons may also be insightful for the larger defense acquisition community.

A few limitations regarding the observations and lessons are worth noting. First, the lessons are primarily based on empirical data from Air Force OT users. Our research approach included collecting and summarizing such lessons but did not explicitly validate them. For this reason, we do not refer to them as "best practice." Further, the lessons we provide were informed primarily from government personnel. We did hold discussions with one OT consortium manager (and benefited from an informal discussion with one Air Force contractor currently under an OT agreement), but perspectives from industry are largely missing from the lessons we present. Future research that explores whether industry perspectives align with lessons presented here and attempts to determine whether lessons presented here are indeed, best practice (e.g., through comparison with OT outcomes) would further bolster the lessons presented in this report.

For the remainder of this final chapter, we summarize this research by returning to the four questions motivating this research that we identified in Chapter 1.

How Has the Air Force Used OTs?

Congress's expanding and granting permanent authority for DoD to use OTs (along with strong support for innovation among proponents in AF senior leadership) appears to have spurred OT activity within the Air Force. While completely accurate data on OT use are not available,[1] overall trends show significant growth since 2016. FPDS-NG data supplemented with data obtained from SAF/AQC show that the number of OTs awarded increased over time from 2016 through 2018 and more than doubled from 2017 to 2018. Analyzing the data by dollar value produces similar results. For the period spanning January 2016 through December 2018, FPDS-NG data show that the Air Force funded at least $1.7 billion in obligations.[2] Spending increased 49 percent from 2016 to 2017 and nearly 140 percent from 2017 to 2018. While a

---

[1] Reporting of OTs does not conform well to the FPDS-NG and therefore is not reported consistently (Schwarz and Peters, 2019).

[2] Given the issues noted with FPDS-NG, we assume this is a lower-bound estimate. FPDS-NG data pulled as of March 7, 2019, and includes Air Force–funded OTs awarded by other services (e.g., Army) and DIU.

single program, NSSL, dominates this dollar total, it should be noted that the Air Force has increased the use of OTs for both single OTs and consortia activities. For example, the demand for OT activities under the SpEC consortium has far exceeded the initial planning basis such that the consortium is in the process of raising its ceiling for the second time. This demand is higher than initially envisioned and suggests that the user community recognizes the value of OT prototypes.

Our case studies suggest that the Air Force has used OTs in a variety of forms and for a number of different types of prototype efforts. Two of our OT project cases were those awarded through consortia; two used cost-sharing; one was a sole-source award; two were awarded by the Army on behalf of the Air Force; two used facilitating organizations (e.g., DIU); and one included a follow-on production OT project. The awards ranged in value from $1 million to more than $100 million; their period of performance ranged from less than a year to more than three years. Prototype efforts included experimental tests, weapon and information technology (IT) systems, and physical and business processes.

While the methods to implement an OT must be tailored to the unique problem and circumstances, each of our cases followed a general OT life-cycle process that began with identifying the problem and strategy, culminating with execution and close out of the OT. As outlined in Chapter 3, interim stages included identifying funding; deciding whether to use an OT; making a number of contract-related decisions (e.g., use of an OT consortium, choosing a contracting office); and facilitating market research, source solicitation, evaluation and selection, and agreement development.

What Are the Outcomes Associated with the Use of OTs?

While no objective measures currently exist for many of the outcomes associated with OTs, participants in every one of our case studies stated that OTs provide them with flexibilities that are either not available or are more difficult, timely, or costly to use under the FAR. The flexibilities included (1) providing for more open communication between contractors and the government during the entire OT life cycle; (2) allowing for new, innovative ways to perform market research and publicize solicitations; and (3) the lack of mandated specifications when developing an agreement. They told us that these flexibilities are attracting nontraditional bidders that provide new sources of innovative technologies and ideas. Most of our case study OTs were, indeed, awarded to nontraditional sources or to traditional sources with significant nontraditional participation. Case study participants told us that the close relationships that are formed between the private and public sector because of OTs' flexible nature also allow for the government to obtain better overall prototypes. While not the classic acquisition measures of success, these findings support the assertion that the Air Force is meeting some intermediate outcomes.

While not necessarily intended to speed acquisition, case study participants told us that OTs can be executed in some cases more quickly than FAR-based transactions. Between 2016 and

2018, the average time the Air Force spent from solicitation release to award was 188 days, with the range being about 50 to 350 days. Unsurprisingly, larger (more expensive and potentially more complex) OTs tend to take longer. However, we lack the comparison data to judge these timelines against more traditional approaches. Based on our case studies, most of the OTs used some form of competition in the solicitation process (only one of the seven was a sole source).[3] The completed OTs in our set of case studies either developed a test item or system, performed a test, or prototyped an activity. In a few cases, the prototype OT has led to follow-on development. None of our case studies had traditional FAR-based follow-on production. Beyond that, it is quite difficult to say whether these prototypes were "successful," as there is no objective definition or uniform measure.

What Are the Enduring Lessons from the Use of OTs That Might Be Helpful to Acquisition Professionals?

In Chapters 3 through 5, we presented a number of observations and considerations about OTs illuminated by our case studies. Given that the Air Force is still learning how to use OTs in the most effective manner, these chapters may create a benchmark for many of the lessons the Air Force has learned in its short history of using OTs. We summarize these lessons as follows.

An advantage of an OT is that it provides flexibility not inherent in FAR procurements. Contracting is more commercial-like. OTs allow for more open communication during negotiation and solicitation. They also allow the government to more effectively partner with industry (e.g., provide technical help) during execution. However, PMs should recognize that their roles on OTs may differ from their roles on traditional FAR-based transactions, as they may require more cooperation.

OTs are one mechanism in the contracting "toolbox" and, as with FAR-based transactions, should be used only when appropriate. OTs can provide the government with increased flexibility to solicit sources and award agreements in a way that attracts nontraditional sources and leverages commercial capabilities. Our case studies suggest, however, that these are necessary but not sufficient reasons to use an OT. Other mechanisms, such as the SBIR program and BAAs, may also provide the necessary flexibilities in some cases. Government personnel still need to determine whether OTs are the most effective award instrument for the problem. Some case study participants also emphasized that conducting an OT simply for the sake of conducting an OT is unlikely to yield the best outcomes for the Air Force.

The rules of good government contract and program management still apply to OTs. FAR embodies a number of commonsense principles that may be applied to the OT environment. Documenting key OT decisions is still necessary for traceability and can help to retain

---

[3] The motivation for this competition may be related to the fact that follow-on production OTs can only be sole-sourced if the prototype OT is competed.

institutional knowledge. External stakeholders (e.g., Congress, OSD, GAO) can still exert their influence on an OT. As discussed in the section on "Agreement Development" in Chapter 5, negotiating agreement terms requires knowledge of subjects including IP rights and liability as well as a range of other subjects. Payment milestones are still important and need to be aligned to key deliverables.

OTs can take on a number of forms or execution mechanisms; government personnel should consider the advantages and disadvantages of the various options. Some examples are that OTs can be implemented as standalone agreements or through consortia; as competed or sole-sourced agreements; as single or multiple award agreements; or through Air Force contracting organizations, other DoD contracting offices, or facilitating organizations. Government personnel need to determine, among other things, which form will provide them with access to the best and most effective sources, technical expertise, and OT experience and how they can achieve the most efficiencies and how various stakeholders will bring their interests to bear.

Early and frequent communication among all OT stakeholders is especially important to the effective use of OTs. Early integration of legal and policy personnel can help to anticipate and mitigate potential disagreements. Anticipating high-level interests (e.g., from funding sources or high-level involvement) can improve smooth execution. Requirements holders and industry may have little experience with working in the OT environment (and for nontraditional sources working with the government completely). AOs must take on the role of educator and honest broker. They must set realistic expectations, for example, on how quickly an OT can be implemented. Contractors new to government contracting may not understand the IP rights the government is requesting, while requirements holders accustomed to using FAR-based IP clauses may be too restrictive in cases where they need not be. AOs must work with both communities to help understand IP options and determine the required level for the need. For OTs, government may often partner with industry to a greater extent than is typical on FAR-based approaches. For example, the Air Force may need to provide technical assistance to help the firm complete the objectives of the OT. All these observations suggest that there is a need for broader, more practical education in the Air Force on the proper use and expectations for OTs.

The problem and OT strategy are not static and may require more user engagement during execution than are normally encountered in FAR-based transactions. As with FAR-based acquisitions, the first step in executing an OT is defining the problem. Unlike acquisitions under a traditional development contract, however, our case studies suggest that the flexibility inherent in OTs allows (and in many cases, requires) the prototype concept and design in an acquisition to evolve as information is gathered during execution of the contract following award. During each iteration, it is critical to ensure the solution is cast as a prototype if an OT is to be used. Initially, many participants from the user community may not understand this nuance. Ideally, in defining the need, users and AOs should be careful in not overspecifying the solution (e.g., defining a technical baseline) such that they preclude other innovative approaches that may meet the need.

Cost-sharing can be an important incentive tool for OTs. While cost-sharing is possible with FAR-based transactions, it requires firms to have certified cost-accounting systems (FAR Part 30) that may be too costly and time-consuming for commercial and nontraditional firms. Cost-sharing with OTs does not have these restrictions and can provide the government with important incentives. Cost-sharing is a strong incentive for firms to perform on an OT agreement. By having "skin-in-the-game," the firms (both traditional and nontraditional) have something to lose. And cost-sharing can be used for nontraditional contractors as well.

While OT agreements require tailoring to the unique need and prototype, they can be informed by a number of existing sources. Our research suggests that AOs use a spectrum of approaches to develop an initial OT agreement but that there are two overarching principles. First, agreement terms and conditions can be informed by a number of existing resources, including example OT agreements, the DoD *OT Guide*, and basic contract law. The FAR/DFARs can also be used as references when developing certain relevant clauses for an OT. These resources are not likely to be useful for every part of the OT agreement, leading to the second commonality: each OT agreement should be tailored to the specific details of the prototype and awardee. Successful past experience designing and executing OTs can improve the ability to tailor the agreement appropriately.

Is There Potential to Improve the Effective Use of the OTs? If So, What Changes in Law, Policy, or Surveillance Might Be Required?

Our case studies suggest that the Air Force is achieving some of the intended outcomes of OTs. However, we only reviewed seven of the more than 20 OTs the Air Force has implemented since 2016, and measuring "success" in OT implementation is still inexact. Thus, we are unable to make the definitive generalization that OTs are being used in their "most effective" form. Further, in the course of our research, we did identify a number of challenges to the effective use of OTs. Mitigating these challenges would likely improve OT use. Some mitigations, such as many of the strategies discussed in the previous section, may be implemented by the OT users themselves. In many cases, however, these user strategies are work-arounds. They treat the symptom, not the underlying problem. Here, we summarize OT-specific obstacles facing the Air Force overall and provide higher-level strategy considerations for how to mitigate them.

While the Air Force has been able to leverage OT flexibility, the current OT environment presents a number of challenges for effective OT use. Few rules apply to OTs that are both generalizable and prescriptive. While we present a number of lessons in this report that the Air Force may use to inform its use of OTs in the future, they are not prescriptive, may differ based on unique qualities of the OT, and therefore, cannot be used for compliance. This results in compliance-based training methods being less effective. Establishing institutional OT knowledge is difficult, as Air Force OT experience is still developing, documentation is not required, and staff turnover is somewhat common. This limited institutional knowledge coupled with the ambiguity that exists in the OT statute (e.g., what constitutes a prototype?) leads to some disagreements and rework. Air Force contracting culture is inherently compliance-based and risk averse. Without defined measures, it is difficult for acquisition professionals to know what constitutes a "successful" OT. This results in discomfort with the risk-taking that may be necessary to effectively implement an OT. Finally, AOs are chosen from a pool of the most experienced and creative COs. If OT use continues to increase, this may result in a less-balanced overall acquisition workforce and a shortfall of qualified acquisition professionals to take on the role of AOs.

The Air Force should consider ways of adjusting its environment to provide relevant training, facilitate information sharing, and manage the OT workforce. Training that focuses on problem-solving (e.g., based on applying skills to example case studies) and not on complying with specific rules may help to better prepare AOs to use informed judgment when encountering the many unique OTs being considered. This training can be informed by lessons gathered by the acquisition community while executing OTs. However, capturing those lessons and ensuring they reach practicing AOs (not just those new to OTs) requires facilitation on the part of the Air Force. A number of more formal coordinating mechanisms (e.g., published lists of personnel with specific expertise, centers of excellence, recurring cross-organization meetings) could make it less difficult for the acquisition community to stay up-to-date on evolving OT strategies, and, in turn, facilitate a practice of continuous learning. The coordinating mechanisms could be implemented within the Air Force, but our research suggests that information sharing across the broader defense acquisition community could also be useful. Finally, planning and management of the OT workforce can help prepare the Air Force for the potential, continuing, increased OT demand. Given that effective OT learning tends to mostly occur in experiential settings, formal mentoring or apprenticeship programs could be considered. Further, it is not just AOs who need to be knowledgeable about OTs—legal and policy personnel as well as requirements holders would all benefit from OT experience. Broader workforce planning including strategies to provide these individuals with broader experience (i.e., across acquisition, technical, and legal) may also improve the effective use of OTs.

Leveraging the full potential of adjustments to the OT environment may require a shift in Air Force culture. The effective use of OTs requires taking calculated risks. The benefit of the Air Force gaining the competitive edge over its near-peer adversaries must be part of that calculation. OT personnel need to be rewarded, not punished, for using their sound judgment when deciding whether to take these necessary risks to accomplish this mission. To be successful, senior leaders should continue to strongly support the usage of OTs where appropriate to drive cultural change toward more risk tolerance and agility.

Appendix A. Case Study Methodology

Overview

This appendix discusses the methods we used to identify lessons and observations from the case study portion of this research. We will discuss case selection considerations, limitations, the data collection process, and the analytical approach used to identify lessons and observations from the Air Force's growing use of OTs.

Case Selection

The basis for characterizing user-focused lessons learned are in-depth studies of OT cases. The validity and applicability of those cases depends greatly on how they were selected. In some studies, single critical cases are sufficient to gain understanding. More often, multiple cases, selected with an eye toward aligning or varying different characteristics of each case, are appropriate. In this project, we opted for multiple cases across a single unit of analysis, known as a *holistic, multiple case design* (e.g., Van Evera, 1997; Yin, 2003). The unit of analysis we chose for our case studies was an OT project. Standalone OT projects can contain a single agreement (as is the case for the engine manufacturing case study) or a number of agreements supporting a common OT project (as is the case for the NSSL RPS case study). For OT projects awarded under consortia, we considered the unit of analysis to be an individual OT project as opposed to the entire consortium.

Widest Span of Cases Was Chosen

To balance the desire to have the greatest generalizability with limited data and resource constraints, we ultimately selected seven cases with varying characteristics such as responsible office, obligation amount, period of performance, and less general characteristics (e.g., prototype content and consortium involvement). Table A.1 illustrates this.

We focused on these characteristics because our review of the literature indicated that they all have some causal relationship to the way OTs are solicited, developed, and executed.

Some Limitations Exist

There were two limitations to consider. First, the sample size is small. This was partly due to the limited time and resources available to the research team; this is a common problem with case study designs. More significantly, though, the small sample size is driven by the fact that the eligible population size is also small. Although the Air Force recently has been executing a large number of OTs, which as shown in Figure 2.1 totaled to $1 billion in obligations in 2018, many of them are still being executed, and stakeholders have not yet had time to fully understand the lessons that might be learned from their activities. To compensate for this, we restricted our population to OTs that had at least been awarded by the beginning of our project (October 2018), allowing us to consider 13 standalone OT projects and three OT consortia (each of which had multiple OT projects). Of those, we chose five standalone OT projects and two OT projects awarded under two different consortia. This choice was informed by two factors: projects that had enough data available for analysis and that represented the greatest cross section of characteristics of interest.

Table A.1. OT Project Case Studies for Analysis

Name	Affiliation	Obligation	Period of Performance (Years)	Features
Counter-UAS	AFRL	$2.0M (in-kind)[a]	< 1	Cost share, experimental test
AOC Pathfinder/ Project Kessel Run	AFLCMC, Army, DIU	$21M	1	Prototype, IT
	AFLCMC	$48M	1	Follow-on, IT
NSSL Rocket Propulsion Systems	SMC	$740M	3+	Cost share, system
Tetra–Space Consortium subaward	SMC	$18M	1–2	Consortium project, system
TAP Lab and OBAC, C5 Consortium subaward	SMC, Army	$2.1M	1–2	Consortium project, IT design
Engine manufacturing process	AFRL	$1.2M	1–2	Sole source, physical process
Technology accelerator	AFRL, AFWERX	$2.5M	< 1	Business process

NOTE: [a] The C-UAS award was a zero-dollar transaction but developed an in-kind cost-sharing agreement.

Second, cost characteristics were underrepresented. This limitation is related to the first; namely, a limited population size. OT projects in the population were either relatively low-cost (less than $10 million obligation) or were very high ($100 million obligation range), with few in between.

Given the limited population size, methodological remedies to these shortfalls are not possible. Therefore, it is important to frame these analyses as an early, systematic look at lessons being learned in a rapidly maturing field.

Case Development Process

The goal of each case was to understand the decisionmaking process for each OT. We also examined the OT's contexts, such as the purpose of the program and the policymaking environment as well. Our primary method for collecting data related to decisionmaking was to conduct a series of semi-structured interviews, which often can give clearer insights into the process by which programmatic decisions were made than documents or other archival materials.

To prepare for interviews, the research team reviewed the relevant OT project-solicitation existing open-source and internal Air Force documents on each case to build background knowledge. We also built a standard semi-structured protocol that we used in all our CO interviews.[1] We sent the protocol's main questions in advance to the interviewees to give them time to digest the topics that we sought to discuss. The generic protocol appears at the end of this appendix.

We conducted interviews in person and via telephone. We interviewed a total of 19 individuals across a variety of organizations, as detailed in Table A.2. We also interviewed one consortium management team.

Table A.2. Organizations Interviewed for Case Studies

Organization	Subdivision
AFRL	Directorate of Contracting
	Directed Energy Directorate
	Munitions Directorate
AFLCMC	Propulsion Directorate Consortium Initiative
	Electronic Systems Center
SMC	Directorate of Contracting
	Launch Enterprise Systems Directorate
	SpEC Directorate
	Advanced Systems & Development Directorate
	Remote Sensing Systems Directorate
DIU	Acquisition and Space leadership
AFWERX	Technology accelerators
Army	ACC-NJ, C5 Consortium

The data were then developed into a set of case reports that described the purpose of each OT, the overall acquisition strategy that the OT fit into, and the decisionmaking process during the development and execution of each agreement. These case reports can be found in Appendix B.

Case Analysis Approach

The interviews yielded a rich source of information for analysis. To identify overarching lessons and observations, we used a four-step process where we

- examined each observation made by an interviewee
- determined if there was an implicit lesson being communicated by the observation
- identified similar lessons across cases to determine the extent to which the interview's lesson could be generalized.

The process is detailed in Figure A.1, and the results are described in Chapter 3.

---

[1] This was modified on an individual basis for discussion with noncontracting personnel relevant to each case, such as the contractor representative, requirements holders, and so on.

Figure A.1. Case Analysis Process

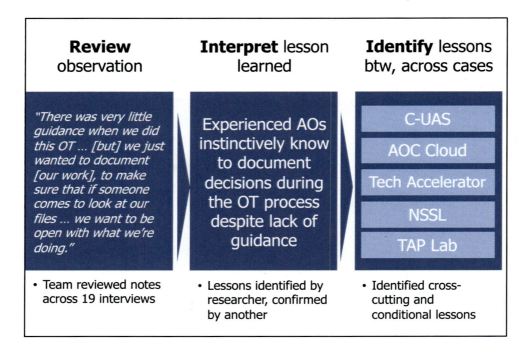

To organize the lessons and observations, we structured them using a framework that focused on themes identified inductively and refined over the course of the data collection.[2] This framework is described in Figure A.2.

Figure A.2. OT Themes for Analysis

By stakeholder involvement	By OT process phase	General considerations
• Service requirements holder/user • Contracting team • Program office • Consortium manager • Contractor • Other DoD agencies & services • Non-DoD stakeholders	• Problem definition • Choosing an OT vs. another mechanism • Identifying funding • Market intelligence & solicitation process • Proposal evaluation and selection • Agreement/SOW development (e.g., T&Cs negotiation) • Execution • Close out	• Price/cost reasonableness • IP and data rights • Resource sharing • Payments • Follow-on activities • Government team oversight • External oversight • Disputes and protests • Consortia • Communication • Training and education • Documentation

NOTE: T&C refers to terms and conditions.

---

[2] We used a number of thematic identification methods described in Ryan and Bernard (2003).

Organizing the lessons and observations according to this framework allows interested practitioners (and policymakers) at all levels to focus on their specific areas of interest in keeping with the theme of this research focusing on user-level insights.

Case Study Protocol

RAND has been asked by SAF/AQC to develop a better understanding of the Air Force's use of other transactions for prototype projects (OTs). We're particularly interested in understanding the best practices and lessons learned that might improve the effective use of OTs in the future.

We will ask a series of questions about your experience with the [name] prototype OT. We specifically want to understand the decisions you made surrounding this OT and what lessons you will use in the future. Any specific vignettes or stories you can contribute that exemplify these key issues would be helpful to our research.

Background Questions

1. What is your position and experience with OTs? Can you briefly describe the OT?
   a. How was the requirement for this capability generated?
   b. How was the OT funded?
   c. What was the deliverable product?
   d. Please describe the roles and responsibilities of others who worked on this OT.

OT Development Questions

2. Why did you decide to use an OT?
   a. What were the alternative approaches, and why were they not chosen?
   b. Why did you choose to (not) use a consortium (vs. doing a standalone award)?
      i. If a consortium was used, why did you choose to use this specific one? Were there other DoD consortia that could have been used/leveraged? If so, why were they not used?
      ii. How was the contracting office/OT authority for this OT chosen?
3. Can you describe how you solicited sources and chose a winner(s)?
   a. What market research did you do?
   b. How did you develop the solicitation, such as selection criteria, process, and so on?
   c. What lessons were learned from the solicitation process?
4. How did you develop the OT agreement?
   a. What knowledge sources (e.g., subject-matter experts, examples, templates, etc.) did you leverage?
   b. How much of the agreement was drafted/proposed by the industry partner?
   c. How did you set up the funding structure?
   d. What were the key sticking points that required negotiation with the contractor, and how were they resolved? How long did the negotiations take?
   e. How did you deal with IP control?

OT Execution Questions

5. During the execution of this OT, did everything go as planned? Why or why not?
   a. How did decisions made during the development of the OT influence the success of its execution?
   b. Were there any terms and conditions negotiated in the OT that turned out to be obstacles?
   c. How did you anticipate and account for modifications during execution?
   d. How did you evaluate the deliverable?
   e. How did the OT conclude?
   f. Did any follow-on or future work result from the OT, or is any planned?

Closing Questions

6. Are there other aspects of this project, or OTs in general, that you would like to share?
7. In hindsight, what good or bad lessons learned—or best practices from your experience with this OT—do you plan to use in the future or document for others?

Appendix B. Case Study Background

This appendix provides background information for our seven case studies, loosely following the structure of the OT life cycle as presented in the main report. Case information includes the purpose of the OT project; the strategy followed to pursue the project; and information about the OT solicitation and agreements, OT awards, and execution of the OT project. Finally, each case concludes with observations and lessons learned from the case.

Case 1. Counter-UAS

A summary of the counter-UAS case study is shown in Table B.1.

Table B.1. Summary Information for the C-UAS Case Study

Funder	Air Force Research Laboratory
Contracting office	Air Force Research Laboratory
Solicitation release date	April 10, 2018
Solicitation response date	May 1, 2018
Date of first award	October 10, 2018
Total government obligation	N/A

Purpose

This project was awarded by the Strategic Development Planning & Experimentation Office (SDPE) at the Air Force Research Laboratory (AFRL). One of SDPE's focus areas is to assess commercially available defense platforms for later transitioning into overseas capabilities. Along these lines, the objective of this OT project was to assess directed energy counter–unmanned aerial systems (C-UAS) technologies in a military-like environment. Rather than contracting with vendors to see a demonstration, this OT enabled Air Force operators to actually test the commercial off-the-shelf (COTS) solutions themselves but without the burden of having to actually purchase any systems outright.

Strategy

For this OT project, SDPE wanted to research and assess C-UAS capabilities but without having to commit to a full system acquisition. SDPE's role was to gather data that could feed into a mid-tier acquisition rather than to perform the acquisition itself. SDPE had designed similar OT agreements in the past, including those for both phases of the Light Attack Experiment (LAE), where vendors brought their COTS systems on-base and trained Air Force personnel how to use them. Rather than pay the vendors directly for their services, however,

SDPE settled on a mutual exchange of in-kind contributions; the vendors benefited from the exposure of their systems in DoD, and DoD benefited from the assessment. In this way, the LAE assessments were completed without any money being transacted.

Following in the footsteps of LAE, SDPE decided to use a similar zero-transaction vehicle for the C-UAS assessments as well. While it may have been possible to contract the C-UAS assessment under a standard FAR-based vehicle, SDPE decided that this rather unique business relationship would be more amenable to an OT.

Solicitation and Agreements

The primary outreach for this OT was through an RFI on FBO on September 15, 2017. The RFI was very detailed, with the intention of conveying the seriousness to potential offerers, encouraging them to participate and ensuring that SDPE got good information from them up front. SDPE also conducted market research and held an industry day on February 9, 2018, to further identify the potential field of offerers. These presolicitation events led up to the release of the Invitation to Participate (ITP) on April 10, 2018.

As with much of this program, the C-UAS ITP was based off of the LAE experience. The ITP laid out a two-step process for source selection. The first step asked offerers to submit preliminary "capability statements," allowing SDPE to see how the proposed solutions compared with its notional requirements. These capability statements were evaluated based on criteria in the ITP, and in the second step, the offerers that were selected to participate were required to submit further detailed information to support the execution of the assessment.

Once the awardees were chosen, negotiations were begun to finalize the agreements. The C-UAS agreements were closely modeled off of the LAE agreements. Since these agreements were zero cost and nothing was actually being procured, a number of clauses that are often contentious, such as IP, were not applicable. Even with the zero cost, to satisfy regulators, SDPE crafted the agreements so as to demonstrate the cost-sharing inherent in the in-kind relationships. Also, since the awardees were bringing equipment on-base for the assessments without remuneration, liability and indemnification were matters of significant negotiation. Working through indemnification took longer than SDPE expected but ultimately did not affect the execution. Air Force operators were pulling the trigger on a high-powered laser without actually procuring it, so SDPE had to find a way to balance risk for both parties.

Awards

Two awards came out of this program, and both of them were awarded to subdivisions of Raytheon (see Table B.2).

Execution

After the agreements were finalized, the assessments went as planned, and everything was successfully completed. The "deliverable" of this OT was an assessment, so there were no

Table B.2. Counter–Unmanned Aerial System OT Agreements

Firm	Government Share[a]	Recipient Share*	Effective Date
Raytheon (NM)	$1.0 million	$1.6 million	October 18, 2018
Raytheon (TX)	$1.0 million	$0.6 million	October 10, 2018

NOTE: [a] In-kind totals. In the case of an OT, in-kind contributions are the value of noncash contributions (i.e., property or services) provided by either the government or contractor.

contract line item numbers or milestones; the main task of the contractors was to just bring their system to the base. The contractors also provided documentation and observations, however, which fed into the final government assessment. There are no plans for a direct follow-on to this OT, but the assessment that came out of it may serve to launch an actual acquisition in the future.

Observations

- In-kind contributions of resources can play the role of cost shares to meet the conditions set in 10 U.S.C. 2371b for conducting an OT with a traditional source.
- One prototype-related OT can serve as an effective model for another, even if both are unique and address different goals. That is, with suitable models available, a prototype-related OT for a unique business relationship need not be tailored from scratch to match that relationship.
- As a corollary, an AO who has designed one prototype-related OT for application to a unique task can use that experience to quickly and confidently design another OT for application in another unique prototype-related task.
- The government can collect detailed operational data of value to DoD on a proprietary system without requiring the transfer of any IP that might be important to the contractor.

Case 2. AOC Pathfinder/Project Kessel Run

A summary of the AOC Pathfinder case study is shown in Table B.3.

Table B.3. Summary Information for the AOC Pathfinder Case Study

Funder	AFRL, 609th AOC, AFMC (Electronic Systems Center), PACAF
Contracting office	ACC-NJ
Solicitation release date	May 1, 2017
Solicitation response date	May 19, 2017
Date of first award	August 4, 2017
Total government obligation	$21 million

Purpose

The AOC cloud software prototype is meant to test ways of transitioning military software applications in the Air Force from purpose-built platforms (known as re-platforming) to cloud-based ones that are more amenable to upgrades, expansions, and other modifications. The prototype is also a first attempt to put into practice several recommendations that the Defense Innovation Board has developed to improve DoD software acquisition (Chappellet-Lanier, 2018).

Strategy

The AOC cloud prototype OT supports the development of a blended project team that combines contractor personnel experienced in software re-platforming with Air Force personnel. The process would first examine an AOC tanker scheduling software that was in need of improvement.[1] The prototype was meant to test the process under the guidance of the contractor personnel before transitioning to an independent Air Force operation in the future.

Solicitation and Agreements

The Air Force solicited this prototype through DIU, using their CSO process. Interested offerers submitted short white papers for review by DIU; the DIU then selected a subset of offerers to make start-up-style "pitches" to the DIU team.[2]

Pivotal Software Inc. ultimately won the CSO. At this point, DIU turned to ACC-NJ to help it develop the OT agreement. An ACC-NJ AO, working in conjunction with DIU and the contractor, Pivotal Software Inc., modified an existing model agreement maintained by DIU. In particular, Pivotal provided input on its software license agreement, helping ACC-NJ develop terms and conditions that were agreeable to both parties. The ACC-NJ team communicated extensively with DIU and Pivotal in the process to ensure that all parties were satisfied with the agreement. A legal team from ACC-NJ also supported the agreement-writing effort by ensuring that all terms and conditions would be legally defensible.

Awards

A single award was made to Pivotal for a period of 12 months. No cost-sharing was involved, as Pivotal was a nontraditional defense contractor at the time of award.

Execution

The AOC cloud prototype is widely considered a success. The blended team (known by its moniker Kessel Run) successfully developed an improved tanker-scheduling software

---

[1] The Defense Innovation Board was particularly interested in this application because its members observed problems with it firsthand during their fact-finding missions.

[2] For further information about the DIU CSO process, see DIU (2016). Note that with the passage of Section 879 of the NDAA for FY2017 and the DFARS OUSD(A&S) Class Deviation, 2018-00016, the use of CSOs to solicit for OT awards may no longer be allowable.

application using proprietary software known as Cloud Foundry, saving the Air Force an estimated $1 million per day of operations (Air Force Technology, 2017; Wallace, 2018).

However, a number of problems occurred during the execution of the agreement that required attention. The agreement milestones sometimes did not align with the payment schedule. Some milestones did not have payments associated with them, making it more difficult for the government to incentivize performance of that milestone. The metrics to measure performance generally were also not specific enough, leaving room for ambiguity as to when the agreement was considered successful.

Nevertheless, the agreement was regarded as a successful prototype. The Air Force, through the Air Force Life-Cycle Management Center (AFLCMC), negotiated a follow-on agreement with Pivotal to re-platform other software applications with the ultimate goal of an independent, Air Force–only re-platforming program. Based on experience from the prototype agreement, AFLCMC determined that the follow-on agreement needed more flexible payment milestones based on labor hours worked rather than performance milestones reached. For an iterative effort like software development, this was a more reasonable approach. AFLCMC continues to administer the follow-on agreement.

Observations

- OTs can be leveraged to prototype not only a physical system but also a method or process.
- IP rights for software-related OTs are a sensitive issue; it is advisable to use the contractor's end user license agreement to understand its starting point in agreement negotiations.
- OTs can address broadly defined needs, not strict military requirements; therefore, government teams should capitalize on the flexibility inherent in OTs and open up lines of communication at all stages of the OT development process to allow for productive iteration with contractors.

Case 3. The Rocket Propulsion System

A summary of the RPS case study is shown in Table B.4.

Table B.4. Summary Information for the RPS Case Study

Funder	SMC
Contracting office	SMC
Solicitation release date	June 2, 2015
Solicitation response date	June 23, 2015
Date of first award	January 13, 2016
Total government obligation	$740 million

Purpose

The RPS OT prototype development is being conducted by SMC. The OT was a part of a broader, major weapon system program: NSSL, formerly known as the Evolved Expendable Launch Vehicle (EELV). This program's objective is to develop a secure, affordable, and reliable space launch capability for DoD and other government payloads. The impetus for the RPS development grew out of a congressional requirement (stated in the FY 2015 NDAA, Section 1604, Public Law 113-291) of the DoD to develop a rocket propulsion system that was made in the United States by 2019. Prior to this point, NSSL space launch relied (in part) on RD-180 engines produced in Russia.

Strategy

As it had done several years prior, SMC's development strategy was to leverage commercial space launch capability and to adapt it to meet the national security missions. The fact that there is a robust commercial launch industry made it a good match to develop technology under an OT agreement. The government team looked at other contractual mechanisms, such as cooperative agreements, technology investment agreements, and FAR-based contracts. They settled on OTs because they wanted to implement cost-sharing. However, cost-sharing under the FAR requires cost-type contracts and government-approved cost-accounting systems. There were several potential companies that did not have that certification or that did not want to pursue it. It is a large overhead function that cannot be justified in a commercial environment.

Partially because of the complexity and partially due to congressional desire, the overall development was broken up into four steps: (1) risk reduction and technical maturation phase aimed at specific technologies, (2) investment in RPS prototype development (the OT that is this case study), (3) Launch Service Agreement (LSA) to prototype a complete launch capability; and finally, (4) competitively awarded launch service contracts. This step approach to an overall development capability turned out to be beneficial for two reasons: Some technologies and commercial developments did not pan out during the OT, thus allowing the Air Force to easily adapt in-stride; and despite Congress delaying funding of Step 3, the program development was able to continue under the RPS OT.

Solicitation and Agreements

SMC used multiple approaches in conducting market research and reaching out to industry: FedBizOpps postings, phone calls, industry days, and internet research. Interestingly, this outreach was successful in getting a private firm, Blue Origin LLC, to reach out to the Air Force and ultimately participate as a development partner on the RPS OTs. Blue Origin's capability was largely unknown, and its identification illustrates the value of doing a broad outreach—even though the industry was thought to be well understood. Also, the market research identified that a point solution (i.e., a form-fit-function replacement for the RD-180) was not the only technical option that would meet the goals of the NSSL program. These alternate approaches helped to shape the solicitation for RPS to encompass a much broader set of developments.

The RPS solicitation was broken into two steps. The first step (which allowed firms just 21 days to respond) was an initial screening to ensure that the proposals met NSSL objectives and that the technologies proposed were mature enough. These initial proposals were kept very short. The second step asked for full proposals from those firms that passed the first step. The requirements were very general to allow for a diversity of approaches. Essentially, the requirements asked for rocket propulsion systems that had a clear link to the overall NSSL launch system objectives. There were four successful awardees (or teams of awardees). Each was granted one or more development projects under a single OT agreement. The individual projects under the four OT awards were quite diverse in funding and scope, reflecting the different focus and commercial baselines for each firm. These projects were incrementally funded over the execution of the OTs due to uncertainty in overall program funding. These differences illustrate the flexibility of OTs to allow for a diversity of technical approaches and adapt quickly to different funding situations.

In terms of negotiation, each OT began with a model agreement proposed by the government with all the parties negotiating from that baseline. The data rights were, generally, one of the challenging areas to negotiate. But another important aspect to the negotiations, recognized in hindsight, was the structuring of milestones and payments. For example, zero-dollar milestones didn't provide any incentive for industry to deliver (e.g., report deliverables). Also, it was important to have enough milestones of the right value to help balance the industry cash flow so that the firms were not having to borrow money. Some milestones were renegotiated in execution to reflect these learnings.

These OTs were executed by the Space and Missile Systems Center, Launch Systems Enterprise Directorate. The solicitation was released on June 2, 2015.

Awards

As described above, there were four OT agreements executed under RPS activities. These agreements are summarized in Table B.5. Another important aspect is that each OT had (initially) at least a one-third cost share from industry—regardless of whether the firm was traditional or not. This level of cost-sharing better aligned industry and government objectives, as all the firms had "skin-in-the-game." Thus, executing the agreement well and on time directly affected each company's bottom line because each had to use some of its discretionary funds.

The effective dates (award dates) were delayed by almost six months due to funding problems.

Execution

Not all the technical developments were successful or were pursued to completion. For example, one technology became commercially unviable during the OT. Both sides were able to quickly adjust the scope and terms of the OT to pursue more fruitful development paths. However,

Table B.5. Rocket Propulsion System OT Agreements

Firm	Gov't Share	Recipient Share	Effective Date	Last Milestone
ATK Launch Systems Inc.	$170M	$120M	January 13, 2016	December 8, 2018
Aerojet Rocketdyne	$340M	$110M	February 29, 2016	December 23, 2019
SpaceX	$98M	$95M	January 13, 2016	December 13, 2018
United Launch Services	$130M	$120M	February 29, 2016	January 1, 2020

the overall developments led to the follow-on series of LSA OTs to produce working system prototypes. These OTs were recently awarded. Another challenge during execution was with payment terms and conditions in the original agreements not being accepted by DCMA/Defense Logistics Agency. These terms had to be revised, but the flexibility of the OT process allowed this to happen quickly.

Observations

- It can be difficult to do cost-sharing with FAR-based transactions. Most commercial firms will not accept the cost-type contract rules needed under FAR to do this (i.e., certified cost accounting, FAR Part 30). Also, these activities are best-effort agreements—there is no guarantee to have any deliveries.
- Congress can influence execution of high-visibility OT projects—it forced NSSL to break the OT execution into steps by isolating RPS (they originally intended to go to a full LSA agreement).
- Dividing the execution of a large development effort into smaller steps is beneficial in reducing risk by putting in break points to adapt strategy. It also allows the companies evolve their solutions incrementally.
- Using a two-step solicitation process can be a useful screening tool (for unresponsive or higher-risk proposals) and allows the government to reconsider if they don't get what they intended.
- It is important that milestones are set up to emphasize things you want to track and/or care about. Avoid having many zero-dollar milestones, as industry partners won't focus on them (e.g., final reports). Milestones should not be vague; rather, milestone completion should be objectively measured. If midcourse changes are necessary, these can be documented in a simple letter signed by both parties.
- An advantage of OTs over traditional FAR approaches is that it is easier to look for and award a diversity of solutions. The aperture is wider than narrower government requirements typically necessary under the FAR. Under OTs it is important to realize that the government may not own the technical baseline.
- Data-rights issues under an OT may be novel to both government and industry. In this case, neither side really understood these well when negotiating.

Case 4. Space Enterprise Consortium/Tetra

A summary of the Space Enterprise Consortium (SpEC)/Tetra case study is shown in Table B.6.

Table B.6. Summary Information for the SpEC/Tetra Case Study

	SpEC (Consortium Manager)	Tetra
Funding agency	SMC	SMC
Contracting office	SMC	SMC
Solicitation release date	June 7, 2017	January 5, 2018
Final proposals due	July 7, 2017	February 9, 2018
Date of first award	November 17, 2017	March 18, 2018
Total government obligation	$214.6 million[a]	$18 million

NOTE: [a] Original award for the consortium was $100 million. Ceiling was increased to $500 million on September 11, 2018.

Consortium Overview

Purpose

SpEC was established by SMC to provide an avenue for nontraditional contractors and commercial entities to efficiently collaborate with the Air Force on opportunities to achieve DoD's joint Space Enterprise Vision. Prior to the establishment of SpEC, Air Force requirements holders interested in pursuing an OT could either pursue a standalone OT, which required finding a warranted AO with relevant knowledge, or use other DoD consortia (e.g., Consortium for Command, Control, and Communications in Cyberspace, or C5) when the technical focus could overlap. No consortium entirely focused on space existed at the time. The impetus for creating SpEC began with the Secretary of the Air Force directing SMC to develop a consortium.

Strategy

Prior to releasing a solicitation for a consortium manager, SMC performed market research consisting of more than a dozen interviews of relevant stakeholders (e.g., consortium managers). In conjunction with this research, SMC developed an acquisition strategy and received approval of a determination and findings for the initial consortium ceiling of $100 million.

Solicitation and Agreements

SMC released an RFI in June 2017 for a consortium manager with very few requirements, allowing offerers the freedom to apply their own approach. Nine offerers submitted proposals. Offers were evaluated using three criteria: prior experience, management approach, and manager compensation. Three offers met the minimum criteria. Among those three, SMC decided that Advanced Technology International (ATI) offered the strongest management approach. ATI is a nonprofit firm that approaches consortium management using a coordinating role. An executive committee acts as a board, in which committee members are voted in by consortium members, and makes funding decisions for the consortium.

The solicitation had included a draft agreement for offerers to provide comments. This agreement was informed by multiple OT consortium manager agreements, including AFRL's Open System Acquisition Initiative (OSAI) and at least one consortium from ACC-NJ. The agreement playing the largest influence was that of the NSSL RPS since it had previously gone through policy and legal review at SMC, and the AOs of SpEC and RPS had an existing relationship. ATI's proposal provided refinements to the agreement. Negotiations included terms and conditions that would flow down to prototype project awards.

Awards

ATI was awarded an agreement with a $100 million ceiling and $0 in funds initially obligated. On awarding prototype projects under the consortium, SMC provides obligated funds to ATI, which then disperses funds to the member obtaining the project award. In September 2018, the ceiling was increased to $500 million. As of May 2019, SMC is seeking another increase in its ceiling to $2 billion.

Execution

As of May 2019, SpEC had made 37 awards across 13 projects, with a total value of $214.6 million. Of these awards, nine have been to nontraditional contractors and four to small businesses. The remaining are almost all cost-sharing agreements. ATI's initial consortium management fee was 6 percent but has been reduced to 3.5 percent as the ceiling has increased. Requirements holders using the consortium range from SMC, the Missile Defense Agency, and AFRL. As OTs become more known and SpEC is marketed to requirements holders, demand has increased.

Thus far, the SpEC program manager has been able to keep up with demand. Some challenges it has experienced include a lack of continuity in personnel, including SpEC program management and legal support. ATI provides support to the SpEC program manager on relevant aspects (e.g., screening proposals) to manage surge periods.

Tetra Project Overview

Purpose

The Tetra project, funded by SMC's Advanced Systems and Development Directorate, sought to acquire prototype small satellites to conduct demonstrations and develop tactics, techniques, and procedures at geosynchronous earth orbit. When the Secretary of the Air Force directed SMC to establish SpEC, the director of SMC's Advanced Systems and Development Directorate chose Tetra to be the consortium's "pathfinder" project.

Strategy

Plans for Tetra began before SpEC was established. Prior to SpEC, SMC issued an RFI through FBO to better understand industry's level of interest in participation and to obtain initial cost information. Tetra was planned as a number of sequential prototype vehicles, in which information from previous versions informs the design of subsequent ones. The first prototype was meant to be more conservative in its approach, with subsequent revisions increasing in risk.

The Tetra requirements holder initially developed a requirements description document that was informed by subject-matter experts and mission partners. The original plan was to make one award for the first phase of the project.

Solicitation and Agreements

The Tetra project was solicited through the SpEC consortia in January 2018. Because of the previous market research through FBO, the government team was able to go straight to the more detailed solicitation, which was an RPP. The government team evaluated offers based on criteria set out in the solicitation and presented the evaluation to the director of SMC's Advanced Systems and Development Directorate. The director believed that two offers were worthy of award and was able to acquire funds to award both through networking.

To develop the agreement, the government team began with the standard boilerplate contractual language as developed by ATI and flowing down from the SpEC OT agreement. Negotiations went relatively smoothly. One offerer required extensive negotiations related to potential liability language.

Awards

Two awards were made in the first phase of the Tetra Project, as shown in Table B.7. Both were nontraditional contractors. Millennium Space System Inc. was acquired by Boeing after the award was made but has been able to keep its nontraditional status because of the organizational structure of the subsidiary.

Table B.7. Tetra OT Agreements

Firm	Government Obligation	Effective Date
Blue Canyon Technologies	$6.7 million	March 18, 2018
Millennium Space Systems	$11.0 million	March 18, 2018

Execution

As of June 2019, the first phase of the Tetra project was ongoing. While there have been a few cost and schedule overruns, the requirements holder has been able to manage the awardees with work-arounds. Both awards required modifications to the agreement because of technical knowledge acquired during the period of performance. A significant amount of information from the two awards is be used to inform the prototype concept for the next phase of the Tetra project.

Observations

- There is concern among the acquisition community about the potential for redundancy in focus among multiple consortia in the future (i.e., two or more OT consortia covering the same domain of potential problems or solutions). If and when this occurs, the industry has to join multiple consortia to remain current on opportunities and may see the market as a "pay-to-play" environment (i.e., in which contractors have to "pay" consortium fees to "play" in the market). Some consortia are prioritizing their parent service/command, and requirements holders see the relationships that exist within their local command as being a large benefit. This may be in part driving consortia redundancy.

- Many times, a requirements holder's rationale for seeking out an OT is because it wants to go fast and/or avoid the FAR. Given that these are not appropriate reasons, AOs act as gatekeepers at times, ensuring that OTs are used for valid reasons and meet the statutory requirements.

- Since OTs are new to many personnel, turnover and changes in OT-related staff create barriers when having to incorporate individuals that don't have experience. Continuity with staff such as consortium program managers (government), legal personnel, and AORs would help to streamline the system.

- Since documentation about OT decisions is not required, there is no common standard or format. This creates challenges. For example, award documentation still needs to go through legal review, possibly resulting in rework and iterations. As another example, technical evaluation by requirements holders still entails some form of documentation to ensure all stakeholders (e.g., users, AOs) concur but can also require iterations for everyone to come to agreement. These documentation challenges are exacerbated by turnover of relevant stakeholders (e.g., legal, AORs).

- Even though the *OT Guide* suggests not starting OT formation using a sample agreement, this is often the current practice. One benefit of using existing samples (especially those from the same command/center) is that they have already passed legal and policy review.

- One means of using OTs is to award them in consecutive phases to build knowledge about a prototype/technology over time and base subsequent solicitations off of information gathered during previous phases.

- A major benefit of OTs is the less restrictive communication the government can have with an offerer and the collaboration that can occur between all stakeholders during negotiation and solicitation. As a result, offerers can have a better understanding of what the government wants, and the government can get a better product.

- Consortium managers consider themselves as having two customers—the government and the consortium members (industry). Managers play the role of educator to both government (to help them learn how make the best use of consortia and OTs) and to industry (to help correct the misconception that working with the government is difficult and to help educate company representatives in how to do business with the government).

Case 5. Tools, Applications, and Processing Lab and Overhead Persistent Infrared Battlespace Awareness Center Prototype Design

A summary of the Tools, Applications, and Processing (TAP) Lab and Overhead Persistent Infrared (OPIR) Battlespace Awareness Center (OBAC) case study is shown in Table B.8.

Purpose

The TAP and OBAC prototype project was funded by SMC to develop a detailed design and physical environment that can expand the use and analysis of remote sensing data collected by

Table B.8. Summary Information for the TAP Lab and OBAC Case Study

Funding agency	SMC
Contracting office	ACC-NJ
Request for white papers release date	June 27, 2017
White papers due	July 18, 2017
Request for prototype project release date	April 3, 2018
Final proposals due	April 17, 2018
Date of first award	May 29, 2018
Total government obligation	$2.1 million

the Space-Based Infrared Systems OPIR to support technical intelligence and battlespace awareness for the warfighters.

The current facility to perform these functions is a leased space from Lockheed Martin. SMC made the strategic decision to develop this capability in a government-owned facility. In 2016, Applied Minds Inc. (AMI) was awarded funding to develop a conceptual design for this facility. The OT prototype project would develop the concept design into a detailed design, construct the facility, and test its capabilities.

Strategy

The requirements holder decided to use an OT to develop the prototype, given its flexibility to craft a targeted solicitation that could reveal whether there were any competitors to AMI for this type of work. The requirements holder also hoped that this approach to targeting would result in a faster process to award.

When this decision to use an OT was made, SpEC did not exist, and there were few AOs within SMC, so the requirements holder pursued other contracting office alternatives for awarding the OT. They sought out the C5 Consortium awarded by ACC-NJ. C5 uses Combat Capabilities Development Command (formally Armament Research Development and Engineering Center) as an OT liaison office to work with requirements holders to develop a statement of need. That statement of need was then passed to ACC-NJ for the solicitation stage.

Solicitation and Agreements

The C5 Consortium posted a request for white papers to its industry members. Three responses were received, and AMI was selected after an evaluation by the requirements holder. Almost a year passed between the request for white papers and the RPP. The long time frame between the two requests was a result of three factors: (1) the need to rescope the SOW to ensure it would be considered a "prototype" according to OT statute, (2) the negotiations with AMI that came about because of the rescoped SOW, and (3) the large number of stakeholders involved in these negotiations.

The original SOW submitted by AMI followed a template provided by C5 and only included developing a detailed blueprint for the TAP Lab and OBAC. After reviewing the SOW, the AO believed that the existing scope would qualify as a prototype. To ensure that the OT would comply with the statute and become a prototype, the AO, requirements holder, and AMI worked together to update a SOW that included two additional elements: building out the design and testing it. In the testing phase, the operators assured the tests of the equipment, and AMI produced a final report. Developing this new SOW required almost daily communication among these three parties as well as the TAP Lab and OBAC user.

AMI voiced major concerns with agreeing to the second and third elements in the SOW, mainly because the results of the first element were unknown. C5 requires OT projects to be awarded a firm-fixed price agreement. Given the unknowns, AMI was hesitant to agree to an up-front price and to the technical details of the second and third elements. It finally agreed to sign on to these elements when they were added as contract options. In this way, both the Air Force and AMI could decide whether to continue after the first element was complete.

When the agreement was finally completed, it included a number of milestone payments associated with scheduled gates (e.g., 30-, 60-, and 90-percent design reviews).

Awards

AMI was awarded an agreement for $2.1 million for the first element of the OT with a period of performance of 11 months.

Execution

During the execution of the first element (developing a detailed blueprint for the TAP Lab and OBAC), the project encountered a few problems. Soon after the award, AMI found that the SOW was not specific enough to provide the proper guidance and requested that it be modified. It had been written generically to allow AMI maximum flexibility, but under a firm-fixed-price (FFP) agreement AMI decided that it needed more certainty to complete the work. Amending the OT agreement was mostly slowed down by the large number of stakeholders that needed to agree to its terms.

A second delay occurred with execution because construction of the facility was delayed. Construction was covered under a completely separate contract, and OT execution was dependent on its successful completion. Once the facility construction had reached a critical point, the OT could finally continue.

Another problem encountered during execution related to personnel turnover at the requirements holder. The OT project from conception to execution has seen four AORs. Each AOR has been new to the OT environment and required time to ascend the learning curve.

One final problem with the execution of this OT involved AMI's lack of understanding surrounding government purpose rights (GPR). The company, a nontraditional contractor, had marked some documents as proprietary that should have been GPR. After educating AMI, this problem was mitigated.

In the end, while the project is now running smoothly, the requirements holder acknowledged that the use of an FFP agreement may increase the cost to the Air Force. The uncertainty surrounding this type of prototype may be better managed under a cost-reimbursement contract.

Observations

- A consortium can be set up to award only FFP-based agreements. Some prototype OTs may not be well suited for FFP-based agreements.
- Ensuring that an RFP/SOW fits the definition of a prototype may require iterations among the requirements holder, contractor, and contracting personnel, and this can be time-consuming and delay OT execution. To determine whether a project is a prototype, requirements holders can ask: What will contractors demonstrate? What does success look like? What are we actually prototyping?
- Providing too much flexibility in the SOW can be a mistake. The SOW should be written with enough specificity that the contractor has confidence that it will be paid for the work executed. Sufficient effort should be exerted to properly write the SOW to avoid amendments to the agreement, which can delay execution. This includes early and frequent communications with the contractor.
- Requirements holders/users and contractors alike are often new to the OT environment and require education and "grooming" to facilitate the successful use of OTs. This is often one of the jobs of contracting personnel (e.g., AOs) and can be time-consuming. When projects are awarded under a consortium, the consortium manager can play this role for the contractors.
- Turnover of relevant OT stakeholders (e.g., AORs) can delay OT execution and reduce efficiency, especially if the stakeholder is new to the OT environment.
- The more stakeholders involved in the execution of an OT, the greater likelihood for delay because of miscommunication and the multiple approvals necessary.

Case 6. Engine Manufacturing Process

A summary of the engine manufacturing process case study is shown in Table B.9.

Table B.9. Summary Information for the Engine Manufacturing Process Case Study

Funder	Air Force Research Laboratory
Contracting office	Air Force Research Laboratory
Solicitation release date	N/A
Solicitation response date	N/A
Date of first award	August 20, 2018
Total government obligation	$1.2 million

Purpose

The purpose of this OT project was to halve the per-unit cost of an existing small engine by supporting an existing contractor with improved processes. Reducing the per-unit cost of the engine was critical to the Air Force's plan to use this engine as part of the Grey Wolf cruise missile program that envisioned employing these missiles in large swarms, requiring a cost-effective engine.

Strategy

The Air Force focused the OT project on improving cost efficiencies for an existing engine manufacturer because that contractor was one of the few makers of the type of engine needed for the Grey Wolf. In fact, both prime contractors for the Grey Wolf (Lockheed Martin and Northrop Grumman) chose the same engine supplier—Technical Direction Incorporation (TDI), one of few companies with the expertise to build the small engines required by the program. Case study participants stated that engines drive about 50 percent of total missile cost. Thus, the fact that TDI was the engine subcontractor for both primes made it easier for AFRL to work directly with the manufacturer.

Additionally, the AFRL contracting office, like other contracting offices across the Air Force, was being encouraged by senior Air Force leaders to find opportunities to use OTs. AFRL also thought it would be beneficial to use an OT to help TDI—a nontraditional defense contractor—work with the government and to enable AFRL to provide TDI with up-front capital to support its work.

Solicitation and Agreements

Because this OT was designed with a specific contractor in mind, it was sole-sourced with no solicitation. However, AFRL did write a "Rational[e] for Sole Source" document, which, while not technically required, documented the sole-source decision in the face of potential oversight.

Because TDI had little experience working with government contracts, the AFRL AO was concerned that TDI might be agreeing to terms that it didn't understand. The AO worked closely with TDI throughout the agreement-writing process to ensure that it fully understood the import to each clause being written. The agreement itself was based largely on an earlier agreement from the OSAI consortium OT and used additional clauses from an AFRL Partnership Intermediary Agreement as well.

Awards

There was a single award as part of this OT as noted above. TDI is a nontraditional defense contractor, so no cost share was required.

Execution

This OT was active as of July 2019 and is progressing without any major obstacles. Initially, AFRL made two modifications to get TDI paid through Wide Area Workflow, ultimately requiring DCMA leadership attention to resolve. The engine unit cost has dropped from $52,000 to $38,000 and is somewhat short of the $26,000-per-unit goal.

Observations

- OTs may be used to test out or try new processes, such as reducing manufacturing costs for an existing product.
- Since many nontraditional defense contractors have (by definition) never worked with the government, AOs may need to be prepared to educate and guide the contractor through the agreements process to ensure that they fully understand the import of each term or condition of the agreement.
- OTs can provide flexibility to work directly with a nontraditional supplier in a unique position (e.g., TDI manufactured the existing engine) and allow the government to support the supplier's unique needs (e.g., needing up-front payments to build working capital).

Case 7. Technology Accelerator

A summary of the Technology Accelerator case study is shown in Table B.10.

Table B.10. Summary Information for the Technology Accelerator Case Study

Funder	Air Force Nuclear Weapons Center, the Defense Threat Reduction Agency, and OSD Nuclear Matters Office
Contracting office	Air Force Research Laboratory (but initially Army Contracting Command—Aberdeen Proving Ground)
Solicitation release date	Unknown
Solicitation response date	Unknown
Date of first award	June 1, 2017
Total government obligation	$2.5 million

Purpose

Start-up accelerators are market enablers that identify and potentially groom start-up firms with ideas that potential users might want to buy or invest in. They seek to make effective matches between start-ups and interested parties quickly and at low cost to the participants. Such accelerators have boomed in the private sector since the early 2000s.[3]

There are a number of technology accelerator models. In the most common one, the accelerator company winnows an open-ended applicant pool of start-ups into a small cohort of start-ups and then mentors and markets to investors or other interested parties. In another approach, a company seeking ideas engages and sponsors an accelerator company to give it exposure to technologies

---

[3] Two of the most well-known startup accelerators, Y Combinator (Y Combinator, 2014) and Techstars (Techstars, 2019), were founded in 2005 and 2006, respectively.

and products from start-ups relevant to its business area. Techstars, for example, has hosted accelerators sponsored by Disney, Comcast, Target, and other major companies. In both of these models, the main elements of the accelerator include a months-long mentoring phase where the start-ups refine their business plans and elevator pitches and then host a "pitch day" at the end, pitching themselves or their products to a public crowd of investors, technologists, and other interested parties.

A group of students at the Air Force Squadron Officer School observed the benefits of start-up accelerator models and advocated for a similar model to support Air Force needs. Some interaction between start-ups and national security customers already existed, but none offered the end-to-end model of a start-up accelerator focused on Air Force needs. The intelligence community's In-Q-Tel makes venture-capital (VC)-like investments in start-ups and provides intensive support for its portfolio companies, but these do not focus on DoD needs. Similarly, the Army Venture Capital Initiative is narrowly focused on Army interests in innovative energy solutions for soldiers, also using OTs to support its operations (Webb et al., 2014). The purpose of this technology accelerator OT was to introduce the Air Force to the accelerator model.

Strategy

The students decided on a C-UAS technology area for their technology accelerator and, with the help of the school commandant, secured funding from the Air Force Nuclear Weapons Center, the Defense Threat Reduction Agency, and the OSD Nuclear Matters Office. The students decided to use an OT, primarily because they thought it would reduce the number of hurdles involved and more easily attract accelerator companies as offerers, and also because they perceived it would be speedier relative to a FAR transaction.

A major relevant consideration in this OT was how the project fits into the definition of a "prototype." The definition in the *OT Guide* at that time stated that prototype projects may include methodologies or processes.[4] The students argued that the technology accelerator was a *business* process and also that contracting with an accelerator company to run it would count as a "novel application of commercial technologies for defense purposes."[5] As described below, OSD-level reviewers thought otherwise, but the Air Force reviewers agreed with the students, and this program was therefore allowed to proceed.

Solicitation and Agreements

The students solicited this OT by taking their Air Force funding to the ACC Aberdeen. Aberdeen had a standing BAA, making it easy to broadcast this OT opportunity. The only things

---

[4] See 2017 *OT Guide* (DPAP, 2017). Note that the new *OT Guide* now clarifies that "a process, including a business process, may be the subject of a prototype project."

[5] See the 2017 *OT Guide*, which includes this phrase in the "prototype" definition.

the students had to do were pay a small processing fee to the Army and write a justification showing that the Army, as well as the Air Force, would benefit from this OT. Aberdeen eventually awarded this OT to Techstars.

After the award, however, a few problems arose. The students running the program had wanted someone senior to announce the award, and they got the Under Secretary of Defense for Acquisition, Technology, and Logistics to agree to this. However, this triggered OSD Public Affairs and General Counsel reviews, and General Counsel decided that an OT was not an appropriate instrument here primarily because of the "prototype" definition, as described above. OSD also worried that the VC nature of this work might be perceived as duplicative of DoD's in-house VC capabilities (i.e., Army Venture Capital Initiative). But while Techstars does have a VC function, the actual contractor here was Powered By Techstars, a separate entity that runs accelerators on contract without any VC function, thereby avoiding the VC conflict of interest. So the Air Force ultimately decided that an OT *was* appropriate in this instance, and AFRL took over the agreement from Aberdeen.

The AOs at AFRL took over the agreement on August 31, 2017—after execution had begun but before the cohort of start-ups had been decided. While AFRL updated the agreement to make it fit its standards, the students reported that it did not change the meaning or intent of the initial agreement.

Awards

The sole awardee under this program was Techstars, as described above. Since Techstars had never worked with the government, the company counted as a nontraditional defense contractor, and therefore no cost share was required.

In addition to this program, however, a "follow-on" was also awarded. The initial OT did not include a clause for follow-on production since the students had not considered this option when the initial agreement was written. However, after this OT was completed, the students reopened the program for another year (and an option year), this time using a FAR-based strategy with General Services Administration contracting. Techstars was awarded this second contract as well.

Execution

The initial phase of the execution mainly involved choosing the cohort of start-ups to participate in the accelerator. Our case study participants stated that this phase was interrupted by the contracting change described above so that while technology accelerators routinely get upward of 1,000 start-up applicants, this one only got around 170. Techstars worked together with the students to down-select to the final ten participants, based on a combination of business potential and military utility. The accelerator and the pitch day at the end were both completed successfully. One start-up, Blind Tiger Communications, received a follow-on award under

10 U.S.C. 2373 to bench-test its product for Special Operations Command, and other participants were apparently also picked up for follow-ons or funding.

Observations

- A commercial business process applying for DoD purposes may count as a prototype.
- Company organization may make a big difference when considering an OT statute. For example, contracting with a particular subsidiary of a company may make it easier to avoid the appearance of duplicative research, something that is prohibited by 10 U.S.C. 2371.

Appendix C. Legal Considerations

Federal law generally requires competition in the selection of awardees for government contracts. "A bid protest is a challenge to the award or proposed award of a contract for the procurement of goods and services or a challenge to the terms of a solicitation for such a contract." (GAO, n.d.). This appendix summarizes the law governing bid protests as it has evolved to treat OT agreements.

The competition requirement in government contracting traces its recent roots to the 1984 CICA (Wittie, 2003). Under CICA, federal agencies must generally "obtain full and open competition through the use of competitive procedures in accordance with the requirements of (statute) and the Federal Acquisition Regulation" when procuring goods or services.[1] Federal procurement law contains several categories of exceptions to this general competitive procurement, including several authorities for "other than full and open competition" (sometimes called "sole-source" procurements) and authorities for commercial purchases or purchases below certain dollar thresholds.

Under the source selection system established by CICA, disappointed bidders[2] may "appeal" a source selection decision in three forums: the agency that made the decision, the Government Accountability Office (GAO), and the COFC. These three forums are largely independent of each other, although the tight timelines of either an agency protest or GAO protest require that they be initiated immediately on a disappointed bidder's learning of an adverse procurement action. Agency protests are the most informal of the three; although the FAR says they should mirror GAO protests to some extent, they are largely conducted at the discretion of the contracting agency. The GAO operates the most prominent and heavily utilized bid protest forum, largely because of its informal arbitration-like procedures and its speed of case resolution

---

[1] Note that CICA's requirements exist both in Title 41 and Title 10 of the U.S. Code—the titles governing civilian agencies and defense agencies respectively. These statutes generally mirror each other except for differences regarding such issues as protest thresholds and the applicability of certain exceptions for competitive procurement. See 41 U.S.C. 3301; 10 U.S.C. 2304.

[2] Under GAO's regulations governing bid protests, an entity must be an "interested party" to file a protest, meaning "an actual or prospective bidder or offeror whose direct economic interest would be affected by the award of a contract or by the failure to award a contract" (4 C.F.R. 21.0). Under federal case law as articulated by COFC and the Court of Appeals for the Federal Circuit, the courts have decided that an "interested bidder" with respect to a bid protest of a contract or solicitation is "an actual or prospective bidder or offeror whose direct economic interest would be affected by the award of the contract or by failure to award the contract." See 31 U.S.C. 3551(2); 28 U.S.C. 1491(b)(1); see also *Am. Fed'n of Gov't Employees, AFL-CIO v. United States*, 258 F.3d 1294, 1302 (Fed. Cir. 2001); *Weeks Marine Inc. v. United States*, 575 F.3d 1352, 1361 (Fed. Cir. 2009). Further, for such a party to show that its "direct economic interest" was affected, it must show it would have been "a qualified bidder" for the procurement at issue. See *Myers Investigative & Sec. Servs. Inc. v. United States*, 275 F.3d 1366, 1370-71 (Fed. Cir. 2002).

that tend to reduce litigation costs for private parties (Arena et al., 2018). However, because the GAO is a congressional office, it lacks the formal powers of a federal court, such as the authority to compel discovery or enforce legal judgments. The third forum, the COFC, is a federal court with these authorities, albeit with special jurisdiction that applies more narrowly to federal contracting disputes than the broader jurisdiction of a federal district court.

The jurisdiction of the GAO and COFC regarding federal contracts is fairly well-settled law, although Congress, the GAO, and COFC periodically change the rules regarding their jurisdiction.[3] However, the law governing GAO and COFC jurisdiction with respect to bid protests involving OT agreements is less fully developed. The sections below describe the law as it has recently evolved to shape the jurisdiction of agency contracting officers, the GAO, and COFC regarding OT agreement bid protests.

Agency Protest Jurisdiction

Executive Order 12979, 60 Fed. Reg. 55171 (October 25, 1995), directed federal agencies "engaged in the procurement of supplies and services" to "prescribe administrative procedures for the resolution of protests to the award of their procurement contracts as an alternative to protests in forums outside the procuring agencies." This executive order was implemented through FAR 33.103, which established procedures for "agency procurement protests." Importantly, as with GAO and COFC protests, FAR 33.103 requires that a protester be an "interested party," which FAR 33.101 defines as "an actual or prospective offeror whose direct economic interest would be affected by the award of a contract or by the failure to award a contract." Agency protests are required by FAR 33.103 to be "concise and logically presented to facilitate review" and must contain information establishing their timeliness, the standing of the protester as an interested party, and a "detailed statement of the legal and factual grounds for the protest, to include a description of resulting prejudice to the protester." Agencies are generally required to hold contract awards while deciding agency-level protests and shall "make their best

---

[3] Congress and the FAR Councils have repeatedly changed the jurisdictional rules and thresholds for protests involving task orders under "indefinite-delivery/indefinite-quantity" (ID/IQ) or "multiple award" contracts, which are commonly used by DoD and other agencies to procure large volumes of services or supplies under a precompeted contract vehicle. In 2018, the FAR Councils implemented a provision in the FY 2017 National Defense Authorization Act by publishing a new version of FAR 16.505 that set the GAO's jurisdictional threshold at $25 million for DoD, Coast Guard, and NASA task orders and $10 million for civilian agency task orders. (Separately, there is a statutory bar under the Federal Acquisition and Streamlining Act [FASA] of 1994 to most task order protests at COFC, codified at 41 U.S.C. 4106[f] and 10 U.S.C. 2304c.) This bar includes modifications to ID/IQ contract task orders as well. See *Akira Tech. Inc. v. United States*, COFC No. 19-1160C, Fed. Cl. October 10, 2019. These jurisdictional limits apply to most protest grounds, but interested parties may still file protests with GAO or COFC alleging that a task order improperly changes the scope, period, or maximum value of an ID/IQ contract.

efforts" to resolve such protests within 35 days, producing written decisions that "shall be well-reasoned, and explain the agency position."[4]

Agency protest decisions are not required to be public, and there is no public repository or data warehouse for agency protest decisions across the government or DoD. Likewise, these decisions have no precedential value. Consequently, it is difficult to assess how agencies have treated their jurisdiction under FAR Part 33 with respect to protests of OT agreements. However, the language of Executive Order 12979 and FAR Part 33 suggests that agencies likely do not have jurisdiction over OT agreement protests. Executive Order 12979's operant section applies to agencies engaged in the "procurement of supplies and services," and specifically to "procurement contracts" by these agencies. This language arguably excludes OT agreements, which as set forth below, have been defined by both GAO and COFC to not be "procurement contracts." Similarly, FAR 33.101 defines a "protest" as a written objection by an interested party to, inter alia, the "solicitation or other request by an agency for offers for a contract for the procurement of property or services" or the "award or proposed award of the contract." Under well-settled GAO and COFC precedent, OT agreements do not fall within these terms.

GAO Jurisdiction

Under federal bid protest regulations, codified at 4 C.F.R. Part 21, the GAO can only consider a bid protest of "a contract for the procurement of property or services." The federal statutes (10 U.S.C. 2371 and 2371b) conferring OT authority on DoD define these agreements as transactions "other than contracts, cooperative agreements, and grants." The GAO's jurisdictional regulations suggest that this language excludes most issues from GAO consideration during bid protests, limiting the scope of GAO bid protest jurisdiction to whether "an agency is improperly using a non-procurement instrument [such as an OT Agreement] to procure goods or services." (4 C.F.R. 21.5) In other words, the GAO may consider whether the agency properly used its authority to choose an OT (over some other contract type) but may not consider the government's actual source selection decision or other details relating to the OT agreement award.

There have been a relatively small number of protests to GAO under these OT statutes and regulations,[5] potentially reflecting some cognizance among the private bar regarding the difficulty of bringing such protests. Table C.1 summarizes the outcomes for these protests.

---

[4] In addition to the FAR, Air Force Policy Directive 51-12 and Air Force Federal Acquisition Regulation Supplement (AFFARS) Part 5333 set forth procedures for alternative dispute resolution (ADR) and agency protests and disputes, respectively. The AFFARS procedural requirements for bid protests mostly describe notification and authority requirements and do not materially change the procedures outlined in FAR 33.103.

[5] There is a parallel body of law governing GAO jurisdiction over cooperative agreements and other nonprocurement instruments used by agencies. As is the case for OT agreements, the GAO has held that it will generally not review protests regarding the award of cooperative agreements and will only review protests relating to whether an agency is properly using its cooperative agreement authority. See, e.g., *Sprint Commc'ns Co. L.P.*, B-256586, B-256586.2, May 9, 1994. This doctrine antedates CICA and the establishment of the current legal system for competitive government contracting and bid protests. See *Matter of Renewable Energy*, B-203149, June 5, 1981.

Table C.1. GAO Decisions in OT Agreement Protests

GAO Decision	Contracting Agency	Protest Outcome
Energy Conversion Devices Inc., B-260514, June 16, 1995	DoD (ARPA)	Denying protest of consortium award where protester failed to show agency that a "procurement contract" was required, drawing on past cases relating to cooperative agreements and other nonprocurement instruments.
Exploration Partners LLC, B–298804, December 19, 2006	NASA	Denying protest with finding that entering and performing "other transactions" could not be the same as entering and performing procurement contracts, concluding that NASA's use of OT agreements was not "procurement of goods and services," which would be subject to GAO bid protest jurisdiction.
MorphoTrust USA LLC, B–412711, May 16, 2016	DHS (TSA)	Denying protest that the Transportation Security Administration (TSA) abused its authority by using "'other transaction' agreements for the TSA Pre® Application Expansion Initiative."
Rocketplane Kistler, B–310741, January 28, 2008	NASA	Denying protest where protester argued NASA was obligated to use a procurement contract to obtain R&D services, even when an internal NASA policy stated an OT agreement "may only be used when the Agency objective cannot be accomplished through the use of a procurement contract."
Oracle America Inc., B-416061, May 31, 2018	DoD (Army)	Sustaining protest of a follow-on production transaction under DoD's OT authority where the Army did not comply with the statutory requirements for such a follow-on OT award.
Blade Strategies LLC, B-416752, September 24, 2018	DoD (Army)	Denying protest as untimely where protester failed to raise OT authority issue prior to the deadline for submission of proposals.
ACI Technologies, B-417011, January 17, 2019	DoD	Denying protest inter alia because "developing standards and practices, developing of training materials, or training of military personnel as activities [were not] outside the scope of a prototype project."
MD Helicopters, B-417379, April 4, 2019	DoD	Denies protest by small business of Army decision not to enter OTA, stating that "we generally do not review protests of the award or solicitations" and that GAO jurisdiction is "limited to a timely pre-award protest that an agency is improperly using its other transaction authority to procure goods or services."

SOURCE: RAND research via GAO legal decisions website and Westlaw.

Most of the GAO cases involve pre-award protests of solicitations for an OT agreement, largely because, as illustrated in *Blade Strategies LLC*, B-416752, September 24, 2018, the GAO will dismiss post-award challenges to the use of OT authority as untimely. The notable exception to this pattern is the case of *Oracle America*, where GAO reviewed the noncompetitive award of an OT agreement for follow-on production following a prototyping OT agreement. However, in that case, the protester (Oracle) was unable to file a pre-award protest per se because there was no public competition for the OT agreement at issue. Rather, Oracle filed its protest eight days after learning of the OT agreement's award via the Army's publication of this action on

FedBizOpps, the official government portal for publication of contracting, procurement, and other transactions.

Taken together, the GAO decisions regarding OT agreements have consistently held these transactions to be outside the scope of traditional GAO bid protest jurisdiction. Each decision has affirmed that OT agreements are not "a contract for the procurement of property or services," and therefore the GAO has no jurisdiction under CICA or its own bid protest regulations to entertain protests regarding the awards of these agreements. Rather, the GAO has consistently held that its jurisdiction regarding OT agreements is limited to whether agencies properly used their discretionary OT authorities.

COFC Jurisdiction

The third venue for bid protests is the U.S. Court of Federal Claims (COFC).[6] The COFC is the modern successor to the Court of Claims, which existed from 1855 to 1982, and is regarded under federal law as a court that operates under the auspices of Article I of the U.S. Constitution (the article governing Congress), not Article III (the article governing the federal courts). Congress established the jurisdiction for the Court of Claims and now for COFC in the Tucker Act of 1887, which is now codified at 28 U.S.C. 1491. In 1982, the 1982 Federal Courts Improvement Act changed the Court of Claims' name (which was changed ten years later to COFC) and established an appellate path from COFC to the U.S. Court of Appeals for the Federal Circuit and potentially to the Supreme Court, should the high court grant review of a Federal Circuit opinion. In addition to jurisdiction over government contract matters, COFC today exercises jurisdiction over myriad types of money claims against the government, civilian, and military compensation cases; Fifth Amendment takings cases; and tax refund suits.

Subsection (b)(1) of 28 U.S.C. 1491 sets forth COFC's jurisdiction under the Tucker Act for bid protests of government contracts. This statute confers jurisdiction on COFC to hear "an action by an interested party objecting to a solicitation by a federal agency for bids or proposals for a proposed contract or to a proposed award or the award of a contract or any alleged violation of statute or regulation in connection with a procurement or a proposed procurement." Subsection (b)(2) empowers COFC to award both equitable relief (e.g., stays on contract awards) and monetary relief (limited to bid and proposal costs), and subsection (b)(4) establishes that

---

[6] With the Administrative Dispute Resolution Act of 1996, Congress eliminated the jurisdiction of federal district courts for contract procurement disputes arising after January 1, 2001. See *Advanced Sys. Tech. Inc. v. Barrito*, 2005 U.S. Dist. LEXIS 39703, *14 (D.D.C. 2005), cited by Carpenter and Schwartz (2018). However, there may be some causes of action relating to procurement or OT agreements that may be brought against the government by disappointed bidders, as illustrated by the transfer of the SpaceX challenge from the Court of Federal Claims to the U.S. District Court for the Central District of California. See *Space Exploration Tech. Corp. v. United States*, Case No. 19-742C (Fed. Cl., August 28, 2019).

COFC shall review agency actions against a standard of whether they were "arbitrary, capricious, an abuse of discretion, or otherwise not in accordance with law."[7]

As of now, the COFC has decided one bid protest arising out of an OT agreement: the May 2019 protest by Space Exploration Technologies Corp. ("SpaceX") of the Air Force's decision to use its OT authority for LSAs as part of the NSSL program, the successor to the EELV program, and subsequent Air Force decisions made under that authority. SpaceX first filed an agency protest in December 2018, and also sought to engage the Air Force in alternative dispute resolution over these decisions. In April 2019, the Air Force rejected SpaceX's agency protest with what SpaceX characterized in its COFC complaint as a summary response, giving rise to its COFC complaint in mid-2019.

In its complaint, SpaceX argues that COFC has jurisdiction over its protest under 28 U.S.C. 1491 and that COFC's jurisdiction over "any alleged violation of statute or regulation in connection with a procurement or a proposed procurement" should be read broadly to include "all stages of the process of acquiring property or services." According to SpaceX, COFC's jurisdiction should extend to the Air Force's alleged violations of procurement law in the award of these OT agreements.[8] On June 28, 2019, the government filed a motion to dismiss SpaceX's complaint, asserting that COFC lacks bid protest jurisdiction to hear SpaceX's challenge to a nonprocurement action. "This case involves a challenge to an award of a non-procurement action based on purported violations of non-procurement statutes [and] is without merit," the government's motion argued, continuing that "[COFC] should dismiss SpaceX's complaint because it lacks jurisdiction to entertain this bid protest challenge to a non-procurement action." In addition to the government, LSA recipients Blue Origin LLC, Orbital Sciences Corp., and United Launch Services LLC joined the case as intervenors.

In August 2019, Judge Griggsby of the COFC issued an opinion in *Space Exploration Tech. Corp. v. United States*, Case No. 19-742C (Fed. Cl., August 28, 2019), granting the government's motion to dismiss for lack of subject-matter jurisdiction over the SpaceX protest but also allowing SpaceX to transfer its action to the federal district court sitting in Los Angeles. The COFC agreed with the government that "there can be no genuine dispute that the LSAs at issue in this dispute are not procurement contracts that fall within the purview of this Court's bid protest jurisdiction," applying a prior Federal Circuit decision involving cooperative farming

---

[7] COFC's jurisdictional statute refers to the Administrative Procedure Act section codified at 5 U.S.C. 706 for this standard and incorporates this standard by reference. The "arbitrary, capricious, abuse of discretion, or otherwise not in accordance with law" standard is a widely used standard in federal administrative law.

[8] This legal theory may be premised on a footnote on a prior COFC opinion, *United Launch Servs. LLC v. United States*, 139 Fed. Cl. 664, 669, motion to certify appeal denied, 139 Fed. Cl. 721 (2018), which cited a prominent treatise stating that "when an agency uses its other transactions authority, it need not comply with the procurement statutes [or] the FAR," but it still must "comply with any other statute that applies to contractual transactions in general" (Cibinic, Nash, and Yukins, 2011); see also 10 U.S.C. 2371(a) (granting OT authority to "the Secretary of Defense and the Secretary of each military department").

agreements "to require that it must dismiss a bid protest matter challenging agency decisions that are related to the award of an agreement that is not a procurement contract." However, in a footnote, COFC limited its ruling to the facts of the SpaceX case, stating that "the Court does not reach the issue of whether other transactions generally fall beyond the Court's bid protest jurisdiction under the Tucker Act." This decision does not fundamentally change the law in this area and reaches a similar outcome to the GAO cases involving protests of OT decisions. Notwithstanding the judge's footnote regarding the limitations on this decision, the SpaceX protest may establish precedent for future protests of OT agreement decisions to COFC.

Jurisdiction Over OTA Consortia

A number of agencies have awarded OT agreements to consortia for purposes of administration and management, such as the Army and Air Force's System of Systems Consortium, the Army's C5 Consortium, or Air Force Space and Missile System Center's SpEC. These consortia are generally operated by nonprofit corporations that are selected by DoD or military departments for "master OT agreements," which functions like a large indefinite-delivery, indefinite-quantity (ID/IQ) contract vehicle.[9] Consortia invite members to join and participate and receive DoD work in the form of agreements that are negotiated between the consortium and members with no direct contractual involvement by DoD. Unlike traditional DoD subcontracts, the OTA and consortium structure sharply limits the extent to which government contract requirements may be passed through to consortia members. Any disputes that may arise between consortia and their members are commercial disputes that generally do not involve the government and are generally resolved under commercial law and the terms of the private consortia agreements. And to the extent that a private entity (such as another potential consortium) seeks to challenge the award of an OT agreement to a consortium, the GAO held in *Energy Conversion Devices Inc.*, B-260514 that its jurisdiction was limited solely to whether an agency was "improperly [] using a nonprocurement instrument where a 'procurement contract' is required." The logic of this decision, as well as the legal structure of OT consortia, would also preclude protests by disappointed consortia members regarding all other actions respecting consortia by the government, including but not limited to the award of work to a specific consortium, or direction to use particular consortium members.

---

[9] As noted above in the GAO section, Congress sharply limited the jurisdiction of the GAO with respect to task order protests, allowing them only before GAO and only for civilian agency task orders in excess of $10 million or DoD/NASA/Coast Guard task orders in excess of $25 million. Although consortia "task orders" (which are probably better characterized as commercial contracts) are likely not subject to protest rules per se because they are not procurement contracts within agency, GAO, or COFC jurisdiction, it is worth noting that agencies do not gain a safe harbor for such task orders by using consortia because most task orders under ID/IQ contracts were already outside the scope of bid protest jurisdiction.

Lessons Learned

There have only been a few GAO protests about OTs decided on the merits, but these decisions do offer some considerations for the Air Force. *Oracle America* and *ACI Technologies* in particular provide some precedent for the proper use of OTs. Both of these bid protest decisions establish that definitions in the DoD *OT Guide* may be used to clarify undefined terms in the OT statute. Therefore, when following the OT statute to develop an OT, the Air Force may benefit from adhering to the definitions in the *OT Guide*. *Oracle America* also makes judgments about follow-on production, including that "successful completion" of a prototype project requires completion of all agreement modifications and that follow-on production is only allowed if the prototype project agreement includes a clause calling for this. Finally, when making bid protest decisions, GAO places weight on government's contemporaneous documentation of the rationale behind its decisions. Thus, documenting decisions as they are made (especially those under GAO and COFC jurisdiction, such as pursuing an OT and designating it as "successfully completed") may improve the Air Force's ability to justify its position during industry protests.

References

Air Force Technology, "Pivotal: Innovative Partnership Saves Big on US Air Force Fuel Costs," webpage, November 30, 2017. As of September 10, 2019:
https://www.airforce-technology.com/features/pivotal-innovative-partnership-saves-big-us-air-force-fuel-costs/

Anderson, Frank J. Jr., *A Plan to Accelerate the Transition to Performance-Based Services: Report of the 912(c) Study Group for Review of the Acquisition Training, Processes, and Tools for Services Contracts*, Washington, D.C.: Department of Defense, AF903T1, June 25, 1999.

Arena, Mark V., Brian Persons, Irv Blickstein, Mary E. Chenoweth, Gordon T. Lee, David Luckey, and Abby Schendt, *Assessing Bid Protests of U.S. Department of Defense Procurements: Identifying Issues, Trends, and Drivers*, Santa Monica, Calif.: RAND Corporation, RR-2356-OSD, 2018. As of July 18, 2019:
https://www.rand.org/pubs/research_reports/RR2356.html

Baldwin, Laura H., Frank Camm, and Nancy Y. Moore, *Strategic Sourcing: Measuring and Managing Performance*, Santa Monica, Calif.: RAND Corporation, DB-287-AF, 2000. As of December 19, 2019:
https://www.rand.org/pubs/documented_briefings/DB287.html

Boyd, Aaron, "The Gatekeepers of the Government's Other Transaction Deals," *Nextgov*, April 18, 2018. As of September 5, 2019:
https://www.nextgov.com/cio-briefing/2018/04/gatekeepers-governments-other-transaction-deals/147524/

Bullock, Heidi H., "Delegation of Other Transactions for Prototype Projects Authority," Headquarters Air Force Material Command, May 9, 2018.

Camm, Frank, "Adapting Best Commercial Practices to Defense," in Stuart E. Johnson, Martin C. Libicki, and Gregory F. Treverton, eds., *New Challenges, New Tools for Defense Decisionmaking*, Santa Monica, Calif.: RAND Corporation, MR-1576-RC, 2003, pp. 221–246. As of February 25, 2020:
https://www.rand.org/pubs/monograph_reports/MR1576.html

Camm, Frank, "Using Public-Private Partnerships Successfully in a Federal Setting," in Robert E. Klitgaard and Paul C. Light, eds., *High-Performance Government: Structure, Leadership, Incentives*, Santa Monica, Calif.: RAND Corporation, MG-256-PRGS, 2005, pp. 179–216. As of February 25, 2020:
https://www.rand.org/pubs/monographs/MG256.html

Camm, Frank, "Federal Agencies Can Adapt Best Commercial Practice to Improve Their Acquisition of Services," testimony presented to the Acquisition Advisory Panel on April 21, 2006, Santa Monica, Calif.: RAND Corporation, CT-261, 2006. As of February 25, 2020: https://www.rand.org/pubs/testimonies/CT261.html

Camm, Frank, Irv Blickstein, and Jose Venzor, *Recent Large Service Acquisitions in the Department of Defense: Lessons for the Office of the Secretary of Defense*, Santa Monica, Calif.: RAND Corporation, MG-107-OSD, 2004. As of February 25, 2020: https://www.rand.org/pubs/monographs/MG107.html

Camm, Frank, Jeffrey Drezner, Beth Lachman, and Susan Resetar, *Implementing Proactive Environmental Management: Lessons Learned from Best Commercial Practice*, Santa Monica, Calif.: RAND Corporation, MR-1371-OSD, 2001. As of February 25, 2020: https://www.rand.org/pubs/monograph_reports/MR1371.html

Campbell, Shelia, and Chad Shirley, "Estimating the Long-Term Effects of Federal R&D Spending: CBO's Current Approach and Research Needs," *CBO Blog*, June 21, 2018. As of September 18, 2019: https://www.cbo.gov/publication/54089

Carpenter, David, and Moshe Schwartz, *Government Contract Bid Protests: Analysis of Legal Processes and Recent Developments*, Washington, D.C.: Congressional Research Service, R45080, November 28, 2018. As of September 5, 2019: https://fas.org/sgp/crs/misc/R45080.pdf

Chappellet-Lanier, Tajha, "Defense Innovation Board Unveils 'Ten Commandments of Software,'" *FedScoop*, April 27, 2018. As of September 10, 2019: https://www.fedscoop.com/defense-innovation-board-unveils-ten-commandments-software/

Cibinic, John Jr., Ralph C. Nash Jr., and Christopher R. Yukins, *Formation of Government Contracts*, 4th ed., New York, N.Y.: Walters Kluwer Law & Business, 2011.

Consortium Management Group, "Award/Contract W15QKN-17-9-5555," Picatinny Arsenal, N.J.: Army Contracting Command, April 20, 2017. As of December 18, 2019: https://cmgcorp.org/wp-content/uploads/2016/08/C5-OTA-W15QKN-17-9-5555.pdf

Cornillie, Chris, "Pentagon OTA Spending Could Top $7 Billion in FY 2019," Bloomberg Government, August 8, 2019. As of September 5, 2019: https://about.bgov.com/news/pentagon-ota-spending-could-top-7-billion-in-fy-2019/

DAU—*See* Defense Acquisition University.

Defense Acquisition University, "Contracting Officer Representative," Acquisition Encyclopedia, undated. As of September 18, 2019: https://www.dau.edu/acquipedia/pages/articledetails.aspx#!53

Defense Acquisition University, "Interactive Adaptive Acquisition Framework Tool," Tools Catalog, November 13, 2019. As of December 17, 2019:
https://www.dau.edu/tools/t/Interactive-Adaptive-Acquisition-Framework-Tool

Defense Innovation Unit, homepage, undated. As of September 5, 2019:
https://www.diu.mil

Defense Innovation Unit, *DIUx Commercial Solutions Opening: How-to Guide*, Mountain View, Calif.: DIU, November 30, 2016. As of March 5, 2020:
https://apps.dtic.mil/docs/citations/AD1022451

Department of Defense, Defense Procurement and Acquisition Policy, "Other Transactions Guide for Prototype Projects," version 1.2.0, January 2017.

DIU—*See* Defense Innovation Unit.

DPAP—*See* Department of Defense, Defense Procurement and Acquisition Policy.

Drezner, Jeffrey A., and Robert S. Leonard, *Innovative Development: Global Hawk and DarkStar: Their Advanced Concept Technology Demonstrator Program Experience, Executive Summary*, Santa Monica, Calif.: RAND Corporation, MR-1473-AF, 2002. As of December 18, 2019:
https://www.rand.org/pubs/monograph_reports/MR1473.html

Dunn, Richard L., "Injecting New Ideas and New Approaches in Defense Systems: Are Other Transactions an Answer?" *Proceedings of the Sixth Annual Acquisition Research Symposium of the Naval Postgraduate School*, Vol. II: *Defense Acquisition in Transition*, Monterey, Calif.: Naval Postgraduate School, No. NPS-AM-09-030, May 2009. As of September 5, 2019:
https://apps.dtic.mil/dtic/tr/fulltext/u2/a528191.pdf

Erwin, Sandra, "Air Force Space Consortium Funding Could Grow to $12 Billion Over the Next Decade," Space News, September 3, 2019. As of September 5, 2019:
https://spacenews.com/air-force-space-consortium-funding-could-grow-to-12-billion-over-the-next-decade/

Fike, Gregory J., "Measuring Other Transactions Authority Performance versus Traditional Contracting Performance: A Missing Link to Further Acquisition Reform," *Army Law*, 2009, p. 33.

GAO—*See* U.S. Government Accountability Office.

Hagel, Chuck, Secretary of Defense, "The Defense Innovation Initiative," memorandum to the Department of Defense, Washington, D.C., November 15, 2014. As of September 5, 2019:
https://archive.defense.gov/pubs/OSD013411-14.pdf

Halchin, L. Elaine, *Other Transaction (OT) Authority*, Washington, D.C.: Congressional Research Service, RL34760, July 15, 2011. As of September 10, 2019: https://fas.org/sgp/crs/misc/RL34760.pdf

Kelly, Daniel J., *IP Rights Under NASA and DoD "Other Transaction" Agreements—Inventions and Patents*, Washington, D.C.: McCarter and English, 2018.

Kim, Yool, Guy Weichenberg, Frank Camm, Brian Dougherty, Thomas C. Whitmore, Nicholas Martin, and Badreddine Ahtchi, *Improving Acquisition to Support the Space Enterprise Vision*, Santa Monica, Calif.: RAND Corporation, RR-2626-AF, 2020.

Maucione, Scott, "OTA Contracts are the New Cool Thing in DoD Acquisition," Federal News Network, October 19, 2017. As of March 7, 2020: https://federalnewsnetwork.com/acquisition/2017/10/ota-contracts-are-the-new-cool-thing -in-dod-acquisition/

Mazmanian, Adam, "OTAs Primed for Growth," *FCW*, August 5, 2019. As of September 5, 2019: https://fcw.com/articles/2019/08/05/otas-set-for-growth.aspx?m=1

Mitchell, Billy, "Acquisition Lead Looks to Save Pentagon Tech from the 'Valley of Death,'" *FedScoop*, December 8, 2017. As of September 5, 2019: https://www.fedscoop.com/acquisition-lead-looks-save-pentagon-tech-valley-death/

Moore, Nancy Y, Laura H. Baldwin, Frank Camm, and Cynthia R. Cook, *Implementing Best Purchasing and Supply Management Practices: Lessons from Innovative Commercial Firms*, Santa Monica, Calif.: RAND Corporation, DB-334-AF, 2002. As of February 25, 2020: https://www.rand.org/pubs/documented_briefings/DB334.html

Office of the Secretary of Defense, "Definitions and Requirements for Other Transactions Under Title 10, United Stated Code, Section 2371b," memorandum to Secretaries of the Military Departments, Commanders of the Combatant Commands, Directors of the Defense Agencies, and Directors of the DoD Field Activities, November 20, 2018. As of December 20, 2019: https://aaf.dau.edu/wp-content/uploads/2018/11/Definitions-and-Requirements-for-Other -Transactions-Under-Title-10_USC_S....pdf

Office of the Under Secretary of Defense for Acquisition and Sustainment, *Other Transactions Guide*, November 2018. As of September 10, 2019: https://aaf.dau.edu/ot-guide/

Office of the Under Secretary of Defense for Acquisition, Technology, and Logistics *An Assessment of Cost-Sharing in Other Transactions Agreements for Prototype Projects*, Washington, D.C.: DoD Report to Congress, March 2017. As of March 7, 2020: https://www.acq.osd.mil/dpap/cpic/cp/docs/Assessment%20of%20Cost-Sharing% 20in%20Other%20Transactions%20Agreements%20for%20Prototype%20Projects.pdf

Olson, Mancur, *The Rise and Decline of Nations*, New Haven, Conn.: Yale University Press, 1982.

OSD—*See* Office of the Secretary of Defense.

OUSD(A&S)—*See* Office of the Under Secretary of Defense for Acquisition and Sustainment.

OUSD(AT&L)—*See* Office of the Under Secretary of Defense for Acquisition, Technology, and Logistics.

Pernin, Christopher G., Elliot Axelband, Jeffrey A. Drezner, Brian Barber Dille, John Gordon IV, Bruce Held, K. Scott McMahon, Walter L. Perry, Christopher Rizzi, Akhil R. Shah, Peter A. Wilson, and Jerry M. Sollinger, *Lessons from the Army's Future Combat Systems Program*, Santa Monica, Calif.: RAND Corporation, MG-1206-A, 2012. As of December 10, 2019:
https://www.rand.org/pubs/monographs/MG1206.html

Roper, William, "Assessing Military Acquisition Reform," presentation to the House Armed Services Committee, U.S. House of Representatives, March 7, 2018. As of December 19, 2019:
https://docs.house.gov/meetings/AS/AS00/20180307/106892/HHRG-115-AS00-Wstate-RoperW-20180307.pdf

Ryan, Gery W., and H. Russell Bernard, "Techniques to Identify Themes," *Field Methods*, Vol. 15, No. 1, February 2003, pp. 85–109.

Schwarz, Moshe, and Heidi Peters, *Department of Defense Use of Other Transaction Authority: Background, Analysis, and Issues for Congress*, Washington, D.C.: Congressional Research Service, 2019. As of September 5, 2019:
https://crsreports.congress.gov/product/pdf/R/R45521

Shelbourne, Mallory, "Geurts: Window to Pursue Innovative Acquisition Strategies is 'Closing,'" *Inside Defense*, July 25, 2019. As of September 5, 2019:
https://insidedefense.com/daily-news/geurts-window-pursue-innovative-acquisition-strategies-closing

Sidebottom, Diane M., "Other Transaction Basics," Defense Acquisition University, undated. As of March 7, 2020:
https://www.dau.edu/tools/Documents/Contracting%20Subway%20Map/resources/DAU%20OT%20BASICS.pdf

Smith, Giles K., Jeffrey A. Drezner, and Irving Lachow, *Assessing the Use of "Other Transactions" Authority for Prototype Projects*, Santa Monica, Calif.: RAND Corporation, DB-375-OSD, 2002. As of December 10, 2019:
https://www.rand.org/pubs/documented_briefings/DB375.html

Snyder, Jeanette, email discussion with Lauren Mayer, "Question about agreement officer warrants/training," December 11, 2018.

Techstars, "10 Years of History," webpage, undated. As of December 18, 2019: http://history.techstars.com/

Temin, Tom, "OTAs Change Defense Acquisition for the Duration," Federal News Network, June 3, 2019. As of September 5, 2019: https://federalnewsnetwork.com/tom-temin-commentary/2019/06/otas-change-defense -acquisition-for-the-duration/

U.S. Air Force, "Other Transaction Authority," Office of Transformational Innovation, undated. As of August 8, 2019: https://www.transform.af.mil/Portals/18/documents/OSA/OTA_Brief_Ver%206Apr2016.pdf

U.S. Code, Title 10, Subtitle A, Part IV, Chapter 139, Section 2371, Research Projects: Transactions Other Than Contracts and Grants, December 12, 2017. As of September 5, 2019: https://uscode.house.gov/view.xhtml?req=(title:10%20section:2371%20edition:prelim) %20OR%20(granuleid:USC-prelim-title10-section2371)&f=treesort&edition=prelim&num =0&jumpTo=true

U.S. Code, Title 10, Subtitle A, Part IV, Chapter 139, Section 2371b, Authority of the Department of Defense to Carry Out Certain Prototype Projects, August 13, 2018. As of September 5, 2019: https://uscode.house.gov/view.xhtml?req=(title:10%20section:2371b%20edition:prelim) %20OR%20(granuleid:USC-prelim-title10-section2371b)&f=treesort&edition=prelim&num =0&jumpTo=true

U.S. Code, Title 10, Subtitle A, Part IV, Chapter 139, Section 2373, Procurement for Experimental Purposes, August 13, 2018. As of September 5, 2019: https://uscode.house.gov/view.xhtml?req=(title:10%20section:2373%20edition:prelim) %20OR%20(granuleid:USC-prelim-title10-section2373)&f=treesort&edition=prelim&num =0&jumpTo=true

U.S. Government Accountability Office, *Use of "Other Transaction" Agreements Limited and Mostly for Research and Development Activities*, Washington, D.C.: GAO, GAO-16-209, January 2016. As of September 10, 2019: https://www.gao.gov/assets/680/674534.pdf

U.S. Government Accountability Office, "What Is a Bid Protest?" Bid Protests, Appropriations Law, & Other Legal Work: FAQs, webpage, undated. As of December 20, 2019: https://www.gao.gov/legal/bid-protests/faqs

U.S. House of Representatives, *National Defense Authorization Act for Fiscal Year 2016: Conference Report to Accompany S. 1356*, Washington, D.C., Report 114-92, November, 2015, pp. 700–701. As of September 10, 2019:
https://www.govinfo.gov/content/pkg/CPRT-114JPRT97637/pdf/CPRT-114JPRT97637.pdf

U.S. Senate, *National Defense Authorization Act for Fiscal Year 2018 Report*, Washington, D.C.: Committee on Armed Services, July 10, 2017. As of December 18, 2019:
https://congress.gov/congressional-report/115th-congress/senate-report/125/1?q=%7B%22search%22%3A%5B%22Referred+sequentially%22%5D%7D

Van Evera, Stephen, *Guide to Methods for Students of Political Science*, Ithaca, N.Y.: Cornell University Press, 1997.

Wallace, Mark, "The U.S. Air Force Learned to Code—and Saved the Pentagon Millions," *Fast Company*, July 5, 2018. As of September 5, 2019:
https://www.fastcompany.com/40588729/the-air-force-learned-to-code-and-saved-the-pentagon-millions

Webb, Timothy, Christopher Guo, Jennifer Lamping Lewis, and Daniel Egel, *Venture Capital and Strategic Investment for Developing Government Mission Capabilities*, Santa Monica, Calif.: RAND Corporation, RR-176-OSD, 2014. As of April 2, 2020:
https://www.rand.org/pubs/research_reports/RR176.html

Williams, Lauren C., "More Scrutiny of OTAs in Defense Bill," *FCW*, June 14, 2019. As of September 5, 2019:
https://fcw.com/articles/2019/06/14/ota-oversight-ndaa-williams.aspx

Wilson, James Q., *Bureaucracy: What Government Agencies Do and Why They Do It*, New York, N.Y.: Basic Books, 1989.

Wittie, Patricia H., *Origins and History of Competition Requirements in Federal Government Contracting*, Annapolis, Md.: White Paper for Ninth Annual Federal Procurement Institute, Annapolis, Md., February 2003.

Wood, B. Dan, and Richard W. Waterman, "The Dynamics of Political-Bureaucratic Adaptation," *American Journal of Political Science*, Vol. 37, No. 2, May 1993, pp. 497–528.

Y Combinator, "About Y Combinator," webpage, June 2014. As of December 18, 2019:
https://www.ycombinator.com/about/

Yin, Robert K., *Case Study Research: Design and Methods*, 3rd ed., Thousand Oaks, Calif.: Sage, 2003.